IFIP

SERIES ON COMPUTER GRAPHICS

Editors: J.L. Encarnação G.G. Grinstein

José L. Encarnação
Georges G. Grinstein (Eds.)

Workstations for Experiments

IFIP WG 5.10
International Working Conference
Lowell, MA, USA, July 1989

Springer-Verlag
Berlin Heidelberg New York
London Paris Tokyo
Hong Kong Barcelona

Prof. Dr. José L. Encarnação
Graphische Datenverarbeitung
Wilhelminenstraße 7
D-6100 Darmstadt
FRG

Prof. Dr. Georges G. Grinstein
Graphics Research Laboratory
University of Lowell
Lowell, MA 01854
USA

Library of Congress Cataloging-in-Publication Data
Workstations for experiments / IFIP WG 5.10 international working conference, Lowell, MA, USA, July 1989; José L. Encarnação, Georges G. Grinstein (eds.). p. cm. –
(IFIP series on computer graphics) "Proceedings of the 1989 IFIP WG 5.10 International Working Conference on Workstations for Experiments" – Pref.

ISBN-13: 978-3-642-75905-5 e-ISBN-13: 978-3-642-75903-1
DOI: 10.1007/978-3-642-75903-1

1. Research–Data processing–Congresses. 2. Science–Experiments–Data processing–Congresses. 3. Microcomputer workstations–Congresses. I. Encarnação, José L. II. IFIP WG 5.10. III. IFIP WG 5.10 International Working Conference on Workstations for Experiments (1989: Lowell, Mass.) IV. Series. Q180.55.E4W67 1990 001.4'34–dc20 90-9057 CIP

Softcover seprint of the hardcover 1st edition 1991

45/3140 – Printed on acid free paper

Preface

This first volume of IFIP Series on Computer Graphics presents the proceedings of the IFIP WG 5.10 International Working Conference on Workstations for Experiments which took place in Lowell, USA, in July 1989.

Powerful new technology has been made available to researchers by an increasingly competitive workstation market. Papers from Canada, Japan, Italy, Germany, and the U.S., to name a few of the countries represented in the volume, discuss how workstations are used in experiments and what impact this new technology will have on experiments.

As usual for IFIP workshops, the emphasis in this volume is on the formulation of strategies for future research, the determination of new market areas, and the identification of new areas for workstation research.

This is the first volume of a book series reporting the work of IFIP WG 5.10. The mission of this IFIP working group is to promote, develop and encourage advancement of the field of computer graphics as a basic tool, as an enabling technology, and as an important part of various application areas. IFIP WG 5.10 has the following tasks in its scope:

- to conduct conferences, working conferences, workshops and symposia as deemed appropriate in furthering its mission,
- to maintain close cooperation with other IFIP working groups (e.g., WG 5.2, WG 5.3, WG 5.7) as users of computer graphics,
- to maintain liaison with other appropriate national and international organizations (e.g., ACM-SIGGRAPH, EUROGRAPHICS) and IFIP affiliates, technical committees and working groups in related fields,
- to stimulate and sponsor research investigations and economic and social studies into the various aspects of the topics in its scope.

This new series will publish proceedings and technical reports as well as extended studies on important topics and trends in computer graphics related to the work of IFIP WG 5.10 or produced by the group. It is intended to publish as the next volumes in this series:

Modeling in Computer Graphics
Proceedings of the IFIP WG 5.10 Working Conference
Tokyo, Japan, April 8–12, 1991

Computer Graphics and Education '91
Goals and Directions for the Next Five Years
Proceedings of the IFIP WG 5.10 Working Conference
Barcelona, Spain, April 4–6, 1991

Other topics and events of IFIP WG 5.10 which are under study and planned, and which will later be considered for publication in the series, include:

- harmonization of existing graphics and related standards,
- the next generation of graphics software (successors to GKS, PHIGS, PHIGS+, PEX, CGI, CGM, ...),
- imaging graphics,
- multimedia graphics,
- graphics arts (computer graphics in and for printing and publishing).

The editors hope that this IFIP book series will serve to strongly support and promote computer graphics not only as a discipline but also as a market segment of increasing relevance, for which education and training as well as technical reporting on its applications, technologies and experiences are of fundamental importance.

Darmstadt, Lowell
August 1990

J. Encarnação, G. Grinstein

Table of Contents

4. Experimental Workstations and Environments

1. Introduction

The Impact of Scientific Visualization on Workstation Development

R. D. Bergeron G. G. Grinstein

1. Introduction

The importance of developing tools for the scientist to present and understand very complex data has become well-established as a major need that spans nearly every scientific and engineering discipline [20]. The enormous diversity and complexity of the problem demands research into multi-dimensional data presentation techniques, studies of human perceptual capabilities, and human factors. The computational, graphical, and human interface demands of scientific visualization will be a dominant force in the growing expectations for workstations of the future.

In order to appreciate the demands that scientific visualization places on workstation environments, it is necessary to understand the broad range of applications that can be classified as *scientific visualization*. Similarly, we must understand the directions that research in scientific visualization is taking in order to understand the implications that this will have for workstation development.

2. Classifying Scientific Visualization

The identification of scientific visualization as an independent discipline has developed so rapidly that there is as yet no well accepted rigorous definition of what is encompassed by the field. In general, just about any technique or application which presents scientific data visually has been classified as scientific visualization. This broad definition includes such diverse techniques as spreadsheets, simple x-y plotting, 3d height plots, grey scale medical imaging, volume visualization, flight simulation, and photorealistic rendering. Although such a broad definition is understandable, it complicates any discussion of the impact of "scientific visualization" on workstation development. It is helpful, therefore, to classify scientific visualization applications and techniques. Such a classification can be based on both the *goals* of the visualization and on the nature of the *data* to be visualized.

2.1 Classification by Goal

One major use of advanced visualization techniques is in applications in which a computer model is used to generate the data to be visualized. The model may be the basis of a simulation of a complex system which generates the data, a solution to a finite element analysis problem, or a solution to a partial differential equation, such as might be used to represent a problem in fluid mechanics. The visualization may be aimed at verification, understanding, or interpretation of the model.

A second important type of visualization is aimed at interpreting real-world data acquired via some form of sampling mechanism. The interpretation of earth satellite data and medical imaging data are the two most common examples of this kind of visualization.

The origin of the data, however, does not provide a very useful understanding of visualization needs. A more meaningful classification can be achieved by considering the fundamental goals of the visualization, which can be roughly divided into two categories:

analytical visualization (directed search): we know what we are looking for in the data; visualization helps to determine whether it is there; and
exploratory visualization (undirected search): we do not know what we are looking for; visualization may help us understand the nature of the data by demonstrating patterns in that data [26].

Analytical visualization is used to answer questions such as: does phenomenon X appear in this data? Exploratory visualization helps to answer questions such as: is there anything significant about this data? Both kinds of visualization questions can be asked by researchers using both model-generated data and sampled data. A scientist with model-generated data uses analytical visualization to verify or interpret the model, and exploratory visualization to understand it. A scientist with sampled data, uses analytical visualization to determine if a particular event or phenomenon occurs in the data. For example, a medical diagnostician viewing output of an NMR scan looks for patterns that indicate the presence of a tumor. This process is analytical visualization because the nature of goal (finding the tumor) is well-understood as are the visual cues that verify (or disprove) the goal. Exploratory visualization is needed when the significance of the relationships among the data components is unknown. Such a situation occurs often in dealing with multivariate statistical data bases.

2.2 Classification by Data

The complexity of a visualization problem increases dramatically with the increasing complexity of the data that must be visualized. It is tempting to categorize an application by the *dimensionality* of its data. Unfortunately, however, the term *dimension* has a variety of interpretations in the context of scientific data specification. Many applications have an inherent *geometric* dimensionality such that the data represents information gathered or computed at points in some geometric space. The information itself can have several components, such as temperature, pressure, etc., so we also define the *parameter* or *data* dimensionality. In addition, the data for many applications consists of collections of data acquired or computed for each of a sequence of time steps. In this context, time can be treated as a dimension independent from both the geometry and the data. We call this the *time-sequence* dimension to distinguish it from the potential use of a time value as simply one of the data parameters [1, 2].

Because many applications have an underlying "natural" geometric interpretation that simplifies the presentation of the data, it is especially helpful to classify visualization problems according to their geometric dimension. Applications falling into each of the geometric categories can also have a time-sequence dimension and any data dimension. It is important to realize that the geometry of a problem can be used to determine how the data is presented to a user, but it need not be. In general, however, the presentation is likely to be most effective if the visual presentation corresponds well with the user's intuitive geometric model of the problem.

One-Dimensional Geometry

The simplest example of an application with one geometric dimension is the sampling of data through time for some process. Such sampling data can be easily presented using a one dimensional display and the time-sequence dimension (corresponding very closely to the essential geometry), or the time-sequence dimension can be transformed to a geometric dimension and the data presented as a two-dimensional graph.

Two-Dimensional Geometry

Many applications can be characterized as being geometrically two-dimensional. Some of these do indeed have truly two-dimensional data, while others have taken on an accepted two-dimensional interpretation through many years of tradition and training. Relationships between two variables are easily presented in two-dimensional graphical layouts including bar charts, pie charts, graphs, scatterplots, etc. Appropriate use of color and techniques such as stacked bar charts and multi-function graphs allow the simultaneous presentation of multiple mappings of one variable against a number of others. We have extensive experience in understanding what formats are effective for what kind of data in this context [3, 7, 19, 25, 26]. By using two-dimensional textures, it is possible to increase even further the dimensionality of data presented within a two-dimensional context [12, 21].

Three-Dimensional Geometry

Many common scientific problems arise out of applications in which the domain is fundamentally three-dimensional. Simulations and analyses of problems in fluid mechanics, molecular modeling, structural analysis, medical diagnosis, etc., all share a fundamental characteristic that the data associated with the problem can be presented (and is probably best interpreted) within the context of real world three-dimensional geometry. There may be many variables associated with a given point or object in the three-dimensional world (such as heat, stress, etc.), but the inherent understanding of the problem is based on a three-dimensional model. Visualization applications with an underlying three-dimensional geometry have attracted the most recent attention in the literature [8–10, 14–16, 18]. Although there are significant unresolved issues concerning the effective and efficient presentation of data in this domain, the underlying context helps to structure the presentation in a natural fashion. Some of the problems in this area also give rise of the drive for better lighting models and more realism in computer graphics [11].

No Obvious Geometry

Perhaps the most difficult data visualization problems are those in which there is no obvious inherent usable geometry in the data. This situation occurs most often with multivariate statistical data bases. In some cases, there simply may not be any geometry in the data; in other cases, the available geometry may not be appropriate for the immediate problem. Perhaps there is an inherent geometry, but it is unknown and, in fact, the scientist's problem is to try to find the inherent geometry. Perhaps there is an inherent geometry, but the dimensionality is so high that it cannot be exploited directly in the visualization process. Clearly, the most difficult problems in visualization come from this domain. In particular, it is likely that progress in understanding how to deal with this domain will be based on learning how to impose a "natural" geometric interpretation on problems that do not have one. In this way, we might be able to transfer our perceptual powers from a natural domain to an unnatural one.

3. Techniques for Scientific Visualization

3.1 Visual Presentation Techniques

The complexity and diversity of applications for scientific visualization give rise to an equally diverse set of visualization techniques. Some approaches are aimed at increasing the number of data parameters that can be simultaneously presented [12, 24], while others attempt to increase the geometrical dimensions that can be perceived by such techniques as volume rendering [9, 22, 27] and stereopsis [13, 17]. The principal unifying theme of existing and evolving visual presentation techniques is one of computational cost, most of which must be borne by the workstation. Because of the large volume of data that must be presented and the increasing sophistication of the techniques, workstations will be expected to provide even more extensive general and graphical computations in order to support effective visualization tools.

3.2 Time as a Perceptual Tool

Animation over time is a particularly promising tool for adding real additional perceptual power to the visual presentation of data. This is especially true when the underlying problem has a time-sequence dimension that can be mapped to the animation time sequence. Unfortunately, however, this is a particularly difficult problem because of the massive amount of data required for each step in the time-sequence and the enormous computation that must typically be performed in order to produce each frame of the animation. Even today's most powerful workstations are limited in their ability to transform time-sequences of complex data into real-time animation. Most such animation is still done by computing the frames off-line and then playing them back.

3.3 Sound as a Perceptual Tool

Research has recently begun in using sound as a tool for presenting complex data. Early results are promising and point to the potential ability to increase the dimensionality of data presented to a user by the simultaneous use of visual and auditory channels [4–6, 28, 29]. Recent techniques have been developed to increase substantially the complexity of data that can be presented using sound [23]. Effective multi-dimensional sound presentation also needs substantial computational power and it is possible that such sound capability will be a standard feature in future visualization workstations.

3.4 Perceptual Understanding

To date sophisticated scientific data visualization techniques have been subjected to very little study relating to their *perceptual* effectiveness. In most cases, visualization programs and packages are put together based on "what the hardware and software can do" and "what looks impressive", rather than on "what the scientist can see". There is seldom any research within the computer science community devoted to discovering, whether humans can perceive the data better in one form than another. In particular, a great deal of research is devoted to developing better illumination models and more realistic image generation, but no one has yet demonstrated that realistic illumination models are particularly useful in scientific visualization. We need to develop a model for defining *perceptual dimensionality* that can serve as the basis for studying the effectiveness of the presentation of complex data. Such a model must relate the various dimensionalities of the *presentation* mode (geometric, data, and time-series) with the dimensionality perceived by the human. Extensive experimentation will be needed to implement a testbed based on such a model and then to evaluate data presentation techniques. Future workstations will be crucial tools in such research.

3.5 Data Manipulation and Transformation

Perhaps the most underrated component of scientific visualization is the need to manipulate large amounts of data. A time-sequence solution to a three-dimensional fluid mechanics problem might produce 20–30 parameters at each of 40,000 points in three space for several hundred time steps. The data dimension of this problem is so high, that the scientist must focus attention on some subset of the data. The visualization environment must provide tools to manipulate and transform very large complex data sets. Such manipulation might include modeling components which attempt to identify and codify abstract data objects within the data. Such manipulation is common in two-dimensional image processing applications; analogous functionality might prove effective in higher dimensions.

4. Visualization Environment

4.1 Hardware Model of a Visualization Environment

The progress of scientific visualization research and the effectiveness of the visualization tools that become available to scientists and engineers will be severely limited by the cost and availability of high performance workstations. Exploratory multi-dimensional, multiparametric visualization will be the most demanding of general purpose computing resources. In particular, such an environment needs massive amounts of high-speed storage, very heavy computational power (probably including large scale parallelism for data manipulation and transformation), and very good, fast graphics capabilities. Many other applications will demand even more extensive graphics facilities, probably requiring special purpose graphics processing, for such tasks as realistic rendering including powerful illumination models.

In all cases, the workstation must be highly interactive and respond in real-time even to complex user requests. Animation should be easily invoked and there should be sufficient memory and power to have it be effective for complex data sets.

4.2 Software Model of a Visualization Environment

The software for a visualization environment shares some characteristics with traditional device independent graphics packages. A major function of a device independent graphics package is to transform an application program's graphical specifications into a standard format and then map that presentation to a variety of physical representations (devices). A visualization environment needs to provide a similar transformation, but at a higher level. The visualization environment needs to be able to transform the application's data so that it matches the input required of a variety of *visualization techniques*. Just as a graphics user wants to be able to switch devices at run-time, so the visualization user must be able to switch among a variety of visualization techniques interactively.

At a higher level still, the visualization environment cannot rely upon a single fixed visualization pipeline. In fact, the visualization user must be able to specify an appropriate visualization pipeline interactively and in real-time. This pipeline must include extensive data manipulation facilities in addition to the actual data presentation tools in a context that supports real-time feedback.

Although we have emphasized data visualization so far, *process visualization* is also emerging as a significant tool for scientists. In many cases, the process used to generate and visualize the data can be as important in understanding the data as the data itself. In effect, this is a method for visualizing the execution of the model. Thus, the data visualization process should be specifiable visually and interactively as well as visualized during execution.

5. Conclusions

In summary, the demands of scientific visualization on workstation hardware development follows the classic situation, namely a need for increased computational performance, more memory, more and faster secondary storage. This is not surprising since the requirements and expectations of advanced users always outpace the available hardware. However, visualization needs go beyond the classic resource demands in that new modes of data presentation, such as sound facilities and novel display techniques, require specialized hardware. This hardware should be made available within the workstation. Examples of such hardware support include sound processing and presentation, stereoscopic displays, real-time animation support, video, graphics, and image processing all integrated in a single box.

Such an integrated hardware environment requires an analogous software platform that is not yet available. The software environment must go beyond the traditional user interface (even considering direct manipulation, window-based systems). It must provide general purpose tools for interactive visualization that support the specification of the visualization process as well as a myriad of perception-based visualization techniques.

References

1. Bergeron, R. D. and G. G. Grinstein, A Reference Model for the Visualization of Multi-Dimensional Data, *Proceedings of Eurographics '89*, September 1989, 393–399
2. Bergeron, R. D. and G. G. Grinstein, A Conceptual Model for Interactive Multi-Dimensional Visualization, Submitted for publication
3. Bertin, J., *Semiology of Graphics*, University of Wisconsin Press, Milwaukee, Wis., 1983
4. Bly, S. A., Presenting information in sound, *Proceedings of the CHI '82 Conference on Human Factors in Computer Systems*, 1982, 371–375
5. Buxton, W., Introduction to this special issue on nonspeech audio, *Human-Computer Interaction 4*, 1 (1989), 1–9
6. Buxton, W., The use of non-speech audio at the interface, *CHI '89 Tutorial #10*, 1989, 2.1–2.15
7. Cleveland, W. S., *The Elements of Graphing Data*, Wadsworth Advanced Books and Software, Monterey, Calif., 1980
8. DeFanti, T. A., M. D. Brown and B. H. McCormick, Visualization: Expanding Scientific and Engineering Research Opportunities, *IEEE Computer 22*, 8 (August 1989), 12–25
9. Drebin, R. A., L. Carpenter and P. Hanrahan, Volume Rendering, *Computer Graphics 22*, 4 (July 1988), 65–74
10. Fuchs, H., M. Levoy and S. M. Pizer, Interactive Visualization of 3D Medical Data, *IEEE Computer 22*, 8 (August 1989), 46–51
11. Greenberg, D., Coons Award Lecture, *Comm. ACM 31*, 2 (February 1988), 123–129
12. Grinstein, G. G. and R. M. Picket, Exvis - An Experimental Visualization Environment, *Proceedings of Graphics Interface '89*, June 1989
13. Grotch, S. L., Three-Dimensional and Stereoscopic Graphics for Scientific Data Display and Analysis, *IEEE Comp. Gr. Appl. 3*, 8 (Nov. 1983), 31–43
14. Haber, R. B., Visualization in Engineering Mechanics: Techniques, Systems and Issues, in *SIGGRAPH '88 Course Notes: Visualization Techniques in the Physical Sciences*, ACM SIGGRAPH, August 1988
15. Helman, J. and L. Hesselink, Representation and Display of Vector Field Topology in Fluid Flow Data Sets, *IEEE Computer 22*, 8 (August 1989), 27–36
16. Hibbard, W. and D. Santek, Visualizing Large Data Bases in the Earth Sciences, *IEEE Computer 22*, 8 (August 1989), 53–57
17. Hodges, L. F. and D. F. McAllister, Stereo and Alternating Pair Techniques for Display of Computer-Generated Images, *IEEE Comp. Gr. Appl. 5*, 9 (Sept. 1985), 38–45
18. Long, M. B., K. Lyons and J. K. Lam, Acquisition and Representation of 2D and 3D Data from Turbulent Flows and Flames, *IEEE Computer 22*, 8 (August 1989), 39–45
19. Mackinlay, J., Automating the Design of Graphical Presentations of Relational Information, *ACM Trans. Gr. 5*, 2 (April 1986), 110–141
20. McCormick, B. H., T. A. DeFanti and M. D. Brown, eds., Visualization in Scientific Computing, *Computer Graphics 21*, 6 (Nov. 1987), 1–9
21. Pickett, R. M. and G. G. Grinstein, Iconographics Displays for Visualizing Multidimensional Data, *Proceedings of the 1988 IEEE Conference on Systems, Man and Cybernetics*, Beijing and Shenyang, People's Republic of China, 1988
22. Sabella, P., A Rendering Algorithm for Visualizing 3D Scalar Fields, *Computer Graphics 22*, 4 (July 1988), 51–58
23. Smith, S., R. D. Bergeron and G. G. Grinstein, Stereophonic and Surface Sound Generation for Exploratory Data Analysis, *Proceedings of CHI '90*, April 1990, 125–132
24. Stettner, A. and D. Greenberg, Computer Graphics Visualization for Acoustic Simulation, *Computer Graphic 23*, 3 (July 1989), 195–206

25. Tufte, E. R., *The Visual Display of Quantitative Information*, Graphics Press, Cheshire, CT, 1983
26. Tukey, J. W., *Exploratory Data Analysis*, Addison-Wesley Publishing Co., Reading, MA, 1977
27. Upson, C. and M. Keeler, VBUFFER: Visible Volume Rendering, *Computer Graphics 22*, 4 (July 1988), 59–64
28. Williams, M. G., S. Smith and G. Pecelli, Computer-human interface issues in the design of an intelligent workstation for scientific visualization, SIGCHI Bulletin 21, 4 (April 1990), 44–49
29. Williams, M. G., S. Smith and G. Pecelli, Experimentally driven visual language design: texture perception experiments for iconographic displays, *Proceedings of the IEEE 1989 Visual Languages Workshop,* Rome, Italy

Issues in User Interface

J. F. Koegel

Participants: John Amason, Lyn Bartram, Alex Brown, Naota Inamoto, Sean McElroy, Dr. Enrico Nordelli, Brian Thompson, Marian Williams. Moderator: Dr. John Koegel

A survey of the participants identified a broad list of user interface issues, which were then grouped in the following categories:

- Information Exchange
- Realtime
- Intelligent Interfaces
- Support for Problem Solving

For each topic the main issues were identified and possible directions were suggested. The discussions are summarized in the following sections.

Information Exchange

In a window-based workstation, a user can interact with multiple applications concurrently. Further, with standards such as X11 and NeWS, client applications can be easily distributed across a heterogeneous computer network. An information exchange policy is needed so that applications can share data and communicate without requiring special conversion programs. These applications may or may not have been intended originally to exchange data.

An information exchange policy could be defined in a layered fashion. The lowest level specifies the internal formats for data exchange or storage. The next level is the data representation level – specification of data types and structure. The third level defines the objects and their relationships – the data semantics of the application. The final level is a description language for the information exchange policy itself. At this level applications describe the policy followed by the lower three layers.

PDES is an example of a layered information exchange standard which is intended for CAD applications. Such an approach is needed for applications in general and should be integrated with the interface so that the user can interactively control the information exchange between applications.

Realtime

The category of realtime is meant to have a broader interpretation than the usual association of hard-realtime control systems. Further, some of the concerns about realtime with respect to the user interface overlap with issues of communication and synchronization in distributed systems in general.

At the interface itself, realtime performance satisfies the user who desires the perception of immediate feedback and response from the objects being displayed and manipulated on the screen. Current systems degrade in performance as the number of objects and their complexity increase. This degradation may be exponential due to thrashing effects. The situation is worsened by the requirements of multimedia presentation including video, images, and animation. Part of the solution lies in improvements in hardware technology. Caching techniques used in display servers for supporting distributed applications also improve efficiency.

Standard window-based interfaces allow the user to interact with multiple concurrent processes each running a separate window. Each process can be generating and receiving events asynchronously with respect to the other processes. Input to the interface comes from a variety of devices and can be synchronized by the server before it is passed on to the application. Events and requests from the client applications can also be synchronized at the server, but may be out of sequence from a global perspective because of delays in communication. Techniques for synchronization of the client application (e.g. time stamped events) can be used at the interface to support display of realtime processes.

Intelligent Interfaces

This discussion focused on reaching a consensus on what it means for an interface to be "intelligent". A second topic related to whether the intelligence is in the interface, application, or both.

A list of characteristics of an intelligent interface (Table 1) was proposed. Each type of intelligent behavior can be supported by both the application and the interface. A separation between the application and the interface was viewed as desirable. In this case, the interface supplies the generic capability and the application the domain specific knowledge. This division of labor isn't a strict requirement. Consider as an example the intelligent front-ends implemented for previously existing numerical applications.

The specification of the support needed to provide these characteristics in an interface is not well understood (see [McElroy90], this volume, for a survey of the state-of-the-art). An illustration of the separation between interface and application in providing intelligent behavior is given for the case of providing

Table 1. Example Characteristics of Intelligent Interfaces

Characteristic	Interface Support	Application Support
Adapt	User Model	Application Model
Learn	Interaction History	Extendibility
Anticipate	Interaction History	Backtracking
Assist	Selectable Dialog Modes	Multiple Levels of Problem Solving
Teach	Student Model Tutorial Module	Expert Module
Explain	Rule Translation	Inference Chain
Symbiosis of applications	Open Architecture	Information Exchange

explanation. Assume that the application produces an inference chain based on its rulebase in deriving the answer to a problem. The interface contains the general-purpose facility for translating the inference chain into an English-like sentence.

Support for Problem Solving

This issue concerns problem solving as a generic activity which is cooperatively performed by the user and the computer. The interface can enhance this joint activity by appropriate use of visual representations. The emphasis is on modeling the problem solving *process* rather than just the problem solving *results*. Modeling the process is important because it provides a context for the results and allows the human problem solver to select between alternate approaches.

There is ample evidence to indicate that similar forms of problem solving occur in many different domains. Recognizing this, allows us to deal with the issue of interface support for problem solving at a general level. In what follows, we briefly characterize the problem in more detail.

Modeling problem solving has been a focus of artificial intelligence, which has identified a repertoire of general problem solving methods, including:

Method	Example Representation
Alternatives	Search Graph
Abstractions	Hierarchical Network
Analogies	Isomorphic Graph Structures
Strategies	Search Tree Imposed on Problem Space
Heuristics	Weighted Arcs and/or Nodes
Deduction	Graph Depicting Deduction Chain
Induction	Decision Tree with Weighted Nodes
Constraints	Constraint Network
Planning	Procedure Network

The most common visual representation used in AI literature is a labeled graph. An important question is whether these existing visual representations can be extended to provide more information to the user. One possible approach could be based on recent work in intelligent tutoring systems. Bonar [Bonar88] has proposed the use of *intermediate representations* for bridging the semantic gap between problem solving and programming languages in a tutor for teaching beginning programmers. An example of an intermediate representation for combinatoric problem solving is given in [Koegel89], where the student explores a tree using specific node operators in order to solve combinatorics problems.

A second question is how to model cooperative problem solving. As a starting point for exploring this issue, work describing problem solving in scientific communities [Kornfield81, Kraut88] can be referred to as well as the literature on computer supported cooperative work (CSCW).

Additional Reading

[Bonar88] Bonar, J. Intelligent Tutoring with Intermediate Representations. Proceedings of the International Conference on Intelligent Tutoring Systems, 1988, pp. 25–32

[Koegel89] Koegel, J. F., Lakshmipathy, N., and Schlesinger, J. An Intermediate Representation for Mathematical Problem Solving, in Computer Assisted Learning (ed. H. Maurer), LNCS 360, Springer-Verlag, 1989, pp. 267–281

[Kornfield81] Kornfield, W. A., Hewitt, C. E., The Scientific Community Metaphor. IEEE Tran. on Systems, Man, and Cybernetics 11(1):24–33, 1981

[Kraut88] Kraut, R. E., Galegher, J., and Egido, C., Relationships and Tasks in Scientific Research Collaboration. Human-Computer Interaction 3(1):31–58 (1987–1988)

[McElroy90] McElroy, S., Employing the Intelligent Interface for Scientific Discovery. *This volume*

2. Tools, Techniques and User Interfaces

Employing the Intelligent Interface for Scientific Discovery

S. McElroy

Abstract

This paper presents the fundamental notions underlying the concept of an intelligent interface by the analysis of recent research. An evaluation of the desirability of this form of interface within an environment of scientific experimentation is discussed. Given the thesis that an intelligent interface is in fact desirable, a proposal is made on how to motivate its utilization.

1. Introduction

One noted effect of the recent proliferation of workstation technology is the placement of ample computing power in the hands of persons whose primary discipline is not computer-related. These people require access to the computational power of the machine but generally have neither the time nor the desire to become computer experts. Ultimately, persons such as these would like to interact with the computing device within the realm of their own sometimes specialized domains.

The purpose of this paper is to explore concepts which have been recently investigated that seem to endow the computing device with the intelligence to manage some of this computer-specific expertise. Additionally, some of these techniques may be used to complement the expertise of the user within his or her own domain of knowledge.

This work is a follow-up on the User Interface Subsession of the IFIP WG 5.10, in whose proceedings this article appears. During these informal discussions, an attempt was made to identify those aspects of intelligence which would have a positive impact on scientific utilization of computers. The following list of ideals was proposed:

- Adaptability
- Ability to Learn
- Smartness About Use of Resources
- Ability to Anticipate Actions
- Ability to Assist the User
- Ability to Teach
- Ability to Explain Actions
- Symbiosis of Application

It seemed clear that a fair amount work had already been performed on this subject. That coupled with the fact that no one present claimed any expertise in this area, left the uneasy feeling that these ideals either may not be realizable or may already have been realized. This paper attempts to characterize the work that has been undertaken. Additionally, these ideals are evaluated, in light of recent research, to ascertain at what level of realization they currently exist.

2. What Makes an Interface Intelligent?

According to the literature, what makes an interface intelligent? [Arens 1988], [Benyon 1988], [Croft 1984], [Mason 1988], [Neal 1988], [Norcio 1989], [Rissland 1984], and [Roth 1987] all focus on the kinds of knowledge that the computing device needs if the interface is to be considered intelligent. Norcio and Stanley present perhaps the most concise list of knowledge sources [Norcio 1989]:

1) *knowledge of the user;* that is expertise with the system,
2) *knowledge of the interaction;* that is, modalities [sic] of interaction and dialogue management;
3) *knowledge of the task/domain;* that is, the ultimate purpose of the problem area and its goals, and
4) *knowledge of the system;* that is, the system's characteristics.

Some preliminary work has been done which addresses each of these knowledge sources at least in part from a practical perspective. Mason and Edwards [Mason 1988] have conducted a survey to evaluate qualitatively the effectiveness of systems to carry off an intelligent dialogue. The systems discussed focus on natural language dialogue as a means of presenting an adaptive or intelligent interface. The systems evaluated are: the BBN project [Sidner 1984] whose components have been built but not integrated, KLAUS [Grosz 1983] for which demonstration systems exist (MICROKLAUS and NANOKLAUS), HAM ANS [Hoeppner 1983] which serves as a frontend for several different application systems, and XCALIBUR [Carbonell 1983] which is a natural language front-end to the XSEL expert system. Judging from the examples cited in [Mason 1988] each of these systems does carry off a reasonably intelligent dialogue with the user within the appropriate application domains.

Mason and Edwards rate each of these systems using a classification system quite different from the four types of knowledge given above.

1) *Control:* "what messages are transmitted, when, by whom and to whom during a dialogue" [Mason 1988]
2) *Models:* "models that a system may have of the domain of the dialogue, of its dialogue partner's plans, knowledge or beliefs, or of its own capabilities." [Mason 1988]
3) *Connectivity:* "relationships between messages that create coherence in a dialogue." [Mason 1988]
4) *Modality:* For example, haptic (alphanumeric terminal), visual, diectic, auditory, voice.

5) *Form:* input/output format (e. g. natural/artificial language, iconographic images, sound).
6) *Knowledge Representation and Inferencing Knowledge Acquisition:* "methods used in a system to represent and use knowledge, the capabilities that a system has for representing knowledge of various kinds, and the specific content that a system has." [Mason 1988]
7) External Information Sources/Targets: Data other than that supplied directly by the user or the system.

Other intelligent interface systems include the CUBRICON [Neal 1988] system, the MONITOR [Benyon 1988] system, an electrocenter design assistant [Cohen 1987], a system for automatic construction of user-interface displays, [Arens 1988] and others. Only the HAM-ANS system has been designed for use in a variety of application domains. And even here only a few have been explored, namely hotel reservations, scene description and a fishery database. All the others are in some way constrained by the application.

2.1 Knowledge of the User

All these systems take into asccount at least some aspect of the sources of knowledge described above. Most pay particular attention to modelling some aspect of human cognition itself. The term mental modelling, devising a mechanism that simulates the way an expert goes about problem solving, is usually used in this context. Another related term is user modelling, mechanisms which the machine employs in an attempt to put user interaction into perspective with relation to the current problem being solved.

Much has been written about user or mental modelling. Several different approaches have been examined. One such approach propounded by Moray argues that [Moray 1987]:

An operator's mental model is a homomorphic decomposition of the real system, and is composed of what the operator believes to be quasi-independent subsystems of the real system.

The operator decomposes systems into smaller independent (from the operator's perspective) subsystems as an aid in interacting with them. If we were to consider the set of all possible states of the system then we would have an isomorphic model of the system since one and only one of these sets exist. Moray proposes a system which appears simple but för which the size of this set is quite large to the extent that an operator is unlikely to use it as her model. And yet the operator does interact with the system and manages it quite well. This implies that the opeator is using a homomorphic model [Moray 1987]:

A partial model of the original in the sense that it can be produced from the original by forming an isomorph of a reduced version of the original.

An important implication of this is that although a homomorphic model may be adequate for normal interaction, an isomorphic model cannot be in every case be generated from it. In other words, should it be necessary to get at the specifics of a particular sub-system, it may not be possible once the model has been reduced. This is further complicated by the observation that higher order homomorphs of the system can be composed of other models which are themselves homomorphic models.

Moray also suggests that the method of decomposition or the content of the sub-systems can be predicted through analysis of the system itself or through empirical evidence gained from observation of the system/operator interaction. This being the case, Moray believes an intelligent aid can be created which remembers the points of diversion from the isomorphic model to the homomorphic model and replaces the more homomorphic models with less homomorphic models when a greater level of detail is needed.

The fact that Moray uses the set of states as a representation of these models suggests that the work performed by Wasserman [Wasserman 1985] and Harel [Harel 1988], and recently implemented in a UIMS by Wellner [Wellner 1989], on the use of state transition diagrams (state charts) as a specification for human computer dialog might be extended to specifiy the hierarchical nature of isomorphic or homomorphic user models.

Another common user model is the stereotype. The system has a finite number of stereotypical qualities which categorize the user. As the interaction proceeds, these qualities are gleaned from the dialogue and the user's stereotype is adjusted accordingly. Elaine Rich [Rich 1983] whose work on the Grundy system (which recommends novels to its user), most exemplifies this type of user model. Where many stereotype user models are an attempt to classify a user somewhere on a discrete scale ranging from novice to expert, Grundy assimilates the information necessary to categorize a user on a wide range of qualities. The purpose of this categorization is the recommendation of novels which the system infers to be relevant to persons who fit a given stereotype. This is done not only by stereotyping the user, but by stereotyping the novels as well and then matching novels to users whose stereotypes match in some manner. Natural language dialogue is used as a refinement mechanism.

Kass and Finin in [Kass 1989] describe a General User Modelling System (GUMS). This is a hierarchically organized stereotype system. The attributes of the stereotypes are acquired either implicitly, through the interaction, or they may be an explicit part of the interface such as in Grundy.

Curiously the two systems whose user models are based on stereotypes are also dialogue systems where humans converse with the system. Moray's work is oriented more toward direct manipulation control of the system.

2.2 Knowledge of the Interaction

As noted above, it's possible that the appropriate choice of user model may be more clearly motivated by the interaction metaphor than by any other factor. Natural language interfaces tend to require a user model which can assist in developing the proper context for interpretation. Direct manipulation interfaces lend themselves more easily to state transition analysis simply because these states have at least a reasonable possibility of being enumerated. Neal et al. have developed a system which employs aspects of both worlds by employing [Neal 1988]:

Simultaneous coordinated diectic pointing gestures with simultaneous coordinated natural language in both user input and system-generated output.

This certainly presents a very powerful tool for interaction. The user converses with the system and uses pointing gestures to fortify points of ambiguity.

It seems that natural language interaction is the predominant intelligent interaction metaphor. Few of these systems employ a graphical format which deviates significantly from a text-based display. And even within the systems which employ natural language, none of them have investigated the use of voice input or audio output.

Regarding the potential for intelligently driven graphical formats, Arens et al. [Arens 1988], have proposed an automatic method for defining and generating graphical presentation of data. Their system employs a knowledge base which encodes information regarding display design in both a domain dependent and independent manner. Then the Presentation Designer, given some data to display, applies this knowledge to produce descriptions of the display. Finally device drivers, given the description, create the desired display. These displays can be used both for information presentation and for input solicitation. McKinlay [MacKinlay 1986] has also dealt with the issue of intelligently displaying data based on the characteristics of the data itself. This work is primarily concerned with presentation of graphs.

2.3 Knowledge of the Task/Domain

A certain amount of intelligence is reflected in the planning involved in getting a task accomplished. To do this, the intelligent system needs an understanding of what the actual goal of the interaction is and how best to assist the operator in attaining that goal. Most systems employ explicit knowledge of the domain to guide the planning process. A few systems use techniques implied by user models to guide progress. The assumption is that an understanding of the users's goals can provide a good method for resolving ambiguities or providing assistance when errors are found or information is requested.

Wolz and Kaiser in [Wolz 1988] use a goal-centered approach with the addition of user modelling. They developed GECIE whose application domain is the Berkeley Unix electronic mail system. The system keeps track of the goals the user has accomplished and how they were accomplished. This information coupled with GECIE's world model, are combined to provide appropriate feedback/guidance to the user. The key concept here is that the system needs explicit knowledge of the domain of the application in order to have an understanding of what the user's goals are and how past experience relates to these goals. How these goals come to be known are delegated to other systems so that GECIE expects goal understanding as input. The systems developed by [Wilensky 1984] and [Pollack 1986] were suggested as possible goal understanding systems.

Another similar system is the explanation generation facility described by Moore [Moore 1989]. This system uses a knowledge base and a query/response analyzer to generate a user goal model. This data is used in conjunction with the dialogue history in order to resolve ambiguous user input and to further customize the interaction to more efficiently proceed toward the perceived goal. The designed system, PEA (Program Enhancement Adivsor) assists users in improving the readability, maintainability or efficiency of their programs.

Norcio and Stanley also report on task modelling as a contributor to the intelligent behavior of a system [Norcio 1989]. They point to a number of systems which employ task/domain knowledge in order to make the system itself more efficient [Greenberg 1985] and [Croft 1984]. These systems do not incorporate any model of the user but are concerned purely with modelling the task domain efficiently. A system which runs efficiently is a definite aid to the user.

2.4 Knowledge of the System

Knowing what a system is capable of and how best to utilize these capabilities is essential to an interface that seems intelligent. Nothing betrays the inanity of a system to a user more than for him to have to cope with poorly conceived input and output devices and constructs or to have to wait patiently while the system responds to some trivial task. The system should be capable of configuring itself to make the best use of resources. If the system is composed of a network of computer systems, then some of the computational burden should be offloaded to underutilized systems on the network. Some computing systems have an afinity for particular varieties of work (e.g. graphics, parallel processing, I/O processing, etc.) the interface could be made to understand this and delegate its workload accordingly. The interface should have some concept of the magnitude of resource consumption and it's effect on the system such that system resources can be scheduled for large resource consuming tasks. A few systems have been proposed for network resource sharing. Apollo's Network Computing System and Sun Microsystems' Open Network Computing are examples.

Input and output techniques could also be optimized based upon interaction efficiency and system resource usage optimization. For example, in a natural language interface, if speech input and/or output are features of the system, the interface should take advantage of them, falling back on text only as a final resort. We also need to recognize the benefits and limitations of different input devices. For example, the actual keyboard is generally agreed to be a superior input device compared to a touchscreen keyboard. But more work needs to be done to ascertain when use of a particular system resource is optimal. For example, there are some applications where use of a touchscreen for pointing seems superior to using a mouse or similar locator device and vice versa.

3. How Can Intelligence in an Interface Aid Scientific Discovery?

Clearly any computer system interface that can make interaction and information processing more efficient in general will make the scientist's job easier. But there are properties of scientific discovery which place special emphasis on some of the criteria discussed above. Perhaps more than any other discipline, scientific experimentation, because of the diversity of domains in which the work takes place, implies the use of non-specific domain knowledge. Although the above discussion would seem to suggest that little work has been successfully completed in this area, at least the specification of a task domain for use in the interface is becoming a reality. Even if the computer cannot independently construct a model for an arbitrary domain, the computer's use of a specified model is certainly a step forward.

The work of scientists is often the result of a desire to impose formalism on some specific domain and further, through these formalisms, to uncover principles which unify various domains. Unification of one domain with another might be stated as a potential goal in the analysis of some experimental data. The intelligent interface might be employed to recognize unifying data as it presents it itself. Or it might be used to keep the scientist on the "right" track toward unification (care needs to be taken here as very often unifying principles are the result of tangential thought). Of course, any goal which can be formulated (e.g. "I want to show a correlation between variable A and variable B") could be used in the analysis phase.

Another large portion of the utilization of computing resources is in conducting experiments and data gathering. An intelligent system for managing and accessing this data cannot help but be a boon to scientific experimentation. Picture a system which recognizes the fact that a particular variety of analysis is better suited for being run on a Cray supercomputer on the network than on the local system and which automatically schedules the job.

4. How Do We Get Intelligent Interfaces into the Hands of Scientists?

Obviously ease of use will not be a deterrent to widespread use of intelligent systems. So the primary obstacle will be cost. A system which incorporates all the functionality discussed will require an inordinate amount of computing resourses. Barring any unique and overwhelming discovery, this goal does not appear to be attainable in the short term. But all is not lost! Even if anyone of these systems results in a reliable product, the impact may be quite significant. Beyond that, if some portion of this research works its way into standardized systems such as the Open Look, Motif or other standards, the migration path for the remaining research will at a minimum be well-established. In addition, it would provide a platform upon which differing applications can, by using the same interface, present at least the appearance of symbiosis. Workstation hardware vendors should be particularly interested in this technology since it makes their machinery available to a wider, albeit less computer oriented, public.

5. Conclusion

Looking back at the IFIP WG 5.10 User Interface Subsession ideals introduced earlier, it appears as though most if not all of them are at least realizable. Some of them have been implemented; adaptability, ability to learn, ability to anticipate, and ability to explain actions. And all but the symbiosis of appliclation has at least been studied. It seems fair to say that the field of intelligent interfaces is well into its childhood and does present the potential for improved utilization of both human and computer resources.

The field needs a bit of coherence to pull itself toward adulthood as voiced by Norcio and Stanley [Norcio 1989]:

It appears, however, that no study has attempted to examine these inter-related issues in the context of a unified approach.

This may seem rather a pessimistic attitude but it's relevance is tantamount to the progession of the field. The work in intelligent interfaces needs to move from the laboratory to the field and it will not be well accepted if it is not presented in a unified pragmatic light.

References

[Arens 1988] Arens, Yigal, Lawrence Miller, Stuart C. Shapiro, and Norman K. Sondheimer, "Automatic Construction of User-Interface Displays," *AAAI '88, The Seventh Conference on Artificial Intelligence (St. Paul, MN, August 21–26, 1988)*, vol. 2, pp. 808–813. Morgan Kaufman, San Mateo, CA 1988

[Benyon 1988] Benyon, D. and Dianne Murray, "Experience with Adaptive Interfaces," *The Computer Journal*, vol. 31, no. 5, pp. 465–473, October, 1988

[Carbonell 1983] Carbonell, J.G., W.M. Boggs, and M.L. Mauldin, "The XCALIBUR Project: a Natural Language Interface to Expert Systems," *Proceedings of IJCAI-83*, pp. 653–656, 1983

[Cohen 1987] Cohen, Robert dM., Jerrold H. May, and Harry E. Pople, Jr., *Transactions on Systems, Man and Cybernetics*, vol. SMC-17, no. 2, pp. 240–249, IEEE, March/April 1987

[Croft 1984] Croft, W.B., "The Role of Context and Adaptation in User Interfaces," *International Journal of Man-Machine Studies*, vol. 21, pp. 283–292, Academic Press Limited, London 1984

[Greenberg 1985] Greenberg, S. and I.H. Witten, "Adaptive Personalized Interfaces – a Question of Viability," *Behavior and Information Technology*, vol. 4, no. 1, pp. 21–45, 1985

[Grosz 1983] Grosz, B.J., M.E. Stickel, and Ed., "Research on Interactive Acquisition and Use of Knowledge: Final Report Covering the Period July 3, 1980 to Nov. 30, 1983," SRI Project 1884, Artificial Intelligence Center, Computer Science and Technology Division, SRI International, 1983

[Harel 1988] Harel, D., "On Visual Formalisms," *Communications of the ACM*, vol. 31, no. 5, pp. 514–530, ACM, New York, May 1988.

[Heoppner 1983] Hoeppner, W., T. Christaller, H. Marburger, K. Morik, B. Nebel, M. O'Leary, and W. Wahlster, "Beyond Domain-independence: experience with the Development of a German Language Access System to Highly Diverse Background Systems," *Proceedings of IJCAI-83*, pp. 588–594, Karlsruhe, West Germany, 1983

[Kass 1989] Kass, Robert and Tim Finin, "A General User Modelling Facility," *Proceedings of CHI, 1988 (Austin, Texas, April 30–May 4, 1989)*, pp. 145–150, ACM, New York, 1989

[Mason 1988] Mason, James A. and Jack L. Edwards, "Surveying Projects on Intelligent Dialogue," *International Journal of Man-Machine Studies*, vol. 28, pp. 259–307, Academic Press Limited, London, 1988

[MacKinlay 1986] Mac Kinlay, Jock, "Automating the Design of Graphical Presentations of Relational Information," *ACM Transactions on Graphics*, vol. 5, no. 2, pp. 110–141, ACM, New York, April 1986

[Moore 1989] Moore, Johanna D., "Responding to 'Huh?': Answering Vaguely Articulated Follow-up Questions," *Proceedings of CHI, 1989 (Austin, Texas, April 30–May 4, 1989)*, pp. 91–96, ACM, New York, 1989

[Moray 1987] Moray, Neville, "Intelligent Aids, Mental Models, and the Theory of Machines," *International Journal of Man-Machine Studies*, vol. 27, p. 619, Academic Press Limited, London, 1987

[Neal 1988] Neal, Jeannette G., Zuzana Dobes, Keith E. Bettinger, and Jong S Byoun, "Multimodal References in Human-Computer Dialogue," *AAAI '88 The Seventh Conference on Artificial Intelligence (St. Paul, MN, August 21–26, 1988)*, vol. 2, pp. 819–823, Morgan Kaufman, San Mateo, CA, 1988

[Norcio 1989] Norcio, Anthony F. and Jaki Stanley, "Adaptive Human-Computer Interfaces: A Literature Survey and Perspective," *IEEE Tkransactions on Systems, Man, and Cybernetics*, vol. 19, no. 2, pp. 399–408, IEEE, March/April 1989

[Pollack 1986] Pollack, M., Inferring Domain plans in Question-answering, Ph. D. Thesis, Moore School, University of Pennsylvania, May 1986

[Rich 1983] Rich, Elaine, "Users are Individuals: Individualizing User Models," *International Journal of Man-Machine Studies*, vol. 18, pp. 199–214, Academic Press Limited, London, 1983

[Rissland 1984] Rissland, Edwina L., "Ingredients of Intelligent User Interfaces," *International Journal of Man-Machine Studies*, vol. 18, pp. 199–214, Academic Press Limited, London, 1984

[Roth 1987] Roth, E.M., K.B. Bennett, and D.D. Woods, "Human Interaction with an Intelligent Machine," *International Journal of Man-Machine Studies*, vol. 27, pp. 479–525, Academic Press Limited, London, 1987

[Sidner 1984] Sidner, C., B. Goodman, A. Haas, M. Moser, D. Stallard, and M. Villain, "Research in Knowledge Representation for Natural Language Understanding," Annual Report 1 September 1983 to August 1984, Report No. 54694, Bolt, Breanek & Newman, Inc., Cambridge, MA, 1984

[Wasserman 1985] Wasserman, Anthony I., "Extending State Transition Diagrams for the Specification of Human-Computer Interaction," *IEEE Transactions on Software Engineering*, vol. 11, no. 8, pp. 699–713, IEEE, August 1985

[Wellner 1989] Wellner, Pierre D., "Statemaster: A UIMS based on Statecharts for Prototyping and Target Implementation," *Proceedings of CHI, 1989 (Austin, Texas, April 30–May 4, 1989)*, pp. 177–182, ACM, New York, 1989

[Wilensky 1984] Wilensky, R., Y. Arens, and D. Chin, "Talking to UNIX in English: An Overview of UC," *Communications of the ACM*, vol. 27, no. 6, pp. 574–593, ACM, New York, June 1984

[Wolz 1988] Wolz, Ursula and Gail E. Kaiser, "A Discourse-Based Consultant for Interactive Environments," *Proceedings of the Fourth Conference on Artificial Intelligence Applications (San Diego, CA, 14–18 March 1988*, pp. 28–33, IEEE, Washington, DC, 1988

A Mixed-Mode Man-Machine Interface for Interactive Problem Solving*

P. Dell'Olmo, E. Nardelli, M. Talamo, and P. Vocca**

Abstract

In the context of problem solving application, the multi-media environments offered by scientific and technical workstations may be used to realize an effective high level of cooperation between the expert and the problem solver. In this paper we build a framework for the design and the development of an intelligent man-machine interface that integrates multimedia workstations' tools (3D graphics, animation, combination of voice and sound) for the realization of analogical and synthetical representation of the complex states of the resolution process. Such interface can provide visual presentation of complex concepts and as consequence on-line interactive evaluation and manipulation of search strategies of the resolution process. A three layered schema of interface architecture is presented and various issues on the mappings between layers are discussed. An experimental prototype of the proposed interface on APOLLO 10000 workstation is being developed at IASI in the framework of the ongoing EEC Project PONTIFEX. Moreover experiments on synthetical and analogical representations are carried on in the University of Rome and consequently a new perceptive experimentation methodology is turning out.

1. Introduction

In the Artificial Intelligence field, heuristics are employed to solve complex problems of difficult nature. Such problems are typically characterized by a very high computational complexity (see, for example, [LuS89] for a synthetic presentation and [GaJ79] for a more exhaustive and formal one). The systems used to solve them usually apply heuristic techniques [Nil71, Pea84].

Unfortunately, as a consequence of problem complexity, even this kind of resolution methods require a long time to be completed and may yield only approximate solutions [Pea84]. In fact these resolution methods are designed for classes of problems rather than for a specific problem instance. The more a heuristic takes into account the characteristics of a specific instance of a problem, the better it performs. So, in order to deal with these instances in an efficient way, it is crucial to furnish to the problem solver specific knowledge about the structure and the characteristics of

* This work has been partially supported by ESPRIT-2 EEC Project PONTIFEX.

** P. Vocca is partially supported by "Consorzio per l'Università a Distanza".

the instance on which it is running (see, for example, [NTV89] where a theoretical approach to interactive guidance of heuristics has been studied). However we must consider that sometimes it may not even be possible to know 'a priori' which are such characteristics. Having such knowledge about the specific problem instances it would be possible both to reduce the execution time required for finding a solution and to improve the accuracy of the result obtained [JMW89].

Our basic idea is that the exchange of such knowledge between the expert and the problem-solving system can be performed by means of a high-level man-machine interface that allows an expert user to drive the resolution strategy of the system by using his/her own knowledge of the problem. In this paper we propose a framework for the design and development of advanced man-machine interfaces that allows an expert to cooperate with a problem-solving system during the resolution process. With such interfaces the expert is allowed to manipulate the search strategies directly and, as a consequence, to adapt the strategy to the specific problem instance according to his/her knowledge and to reduce execution time [Voc89, NTV89].

The interaction is made possible by means of a synthetic and analogical representation of the problem instance and of the current state of the resolution process. Such a representation is obtained by exploiting the functionalities of multi-media tools offered by scientific and technical workstations [FaZ84]. Three-dimensional images, animation, sound and voice are integrated and combined by the interface functionalities, providing a complex but synthetic representation of the problem world that allows the expert to understand better and more quickly how the system is working. This complex and synthetic representation allows not only the description of the characteristics of the problem which have an exclusively "physical" nature, (i.e., the problem variables with direct or indirect meaning or interpretation in the physical world such as speed, pressure, temperature, Mach number, etc.) but also those with an abstract or symbolic meaning, such as strategy type, solution loci, solution optimality, etc.

Studies and experiments in this research area are carried on in the ongoing EEC ESPRIT-2 Projects PROMISE (Process Operator's Multi-media Intelligent Support Environment), whose aim is improving the human interface to knowledge based system in process control and real time applications, and PONTIFEX (Planning of Non-specific Transportations by an Intelligent Fleet Expert) where scheduling problems of a commercial fleet of vehicles are considered.

The paper is organized as follows. In Sect. 2 we propose new modalities of cooperation between the expert and the problem-solving system. In Sect. 3 we describe how this cooperation is realized by means of integrated interface multi-media functionalities. In Sect. 4 a schema of the architecture of the proposed environment is presented. Finally, Sect. 5 contains conclusions and indications of further research.

2. A Cooperative Approach for Problem Solving

In this context an expert is considered a person having deep and specialized knowledge about a well-defined application area. A problem solving system is seen as a system which utilizes whatever operators and methods (heuristics) are available to solve a given problem instance of a certain class [NeS72].

Given a problem instance, defined by an initial state and a set of final states, the problem-solving system applies one of the available methods for generating a search graph [Nil71, Pea84], repeatedly choosing a node (representing an intermediate state) to expand and an operator with which to expand it. The choice of the node to be expanded and the way to expand it depends on the method applied [Pea84]. The expert can only wait during this process, as no effective possibilities of cooperating with the system are provided. The resulting kind of interaction between the expert and the problem-solving system is sketched in Fig. 1.

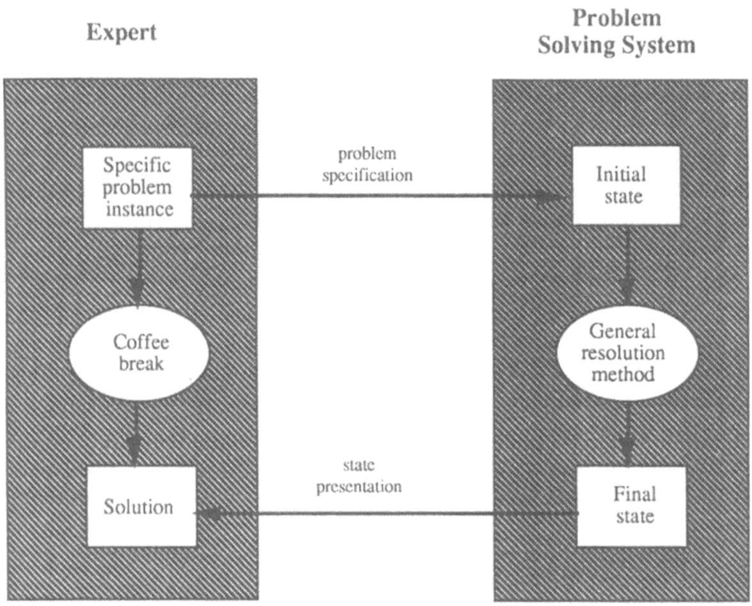

Fig. 1

A problem-solving system is not completely tuned for the specific problem instance due to its limited capabilities in considering novel situations. Hence is not able to adapt the search strategy to coping optimally with the problem instance. On the other side, an expert can utilize past experiences or his/her creativity while solving a problem. This leads to an integrated human-computer problem solver.

If a problem-solving system is provided with tunable methods [Voc89, NTV89], the expert can drive the problem-solving system to get the strategies closest to the characteristics of the problem instance. In order to reach an effective tuning, the expert should be able to recognize the state of the problem-solving system (i.e., the particular problem configuration, the behavior of the resolution method). Such information is represented at system level as complex data structures, algorithms, and heuristics. The goal we want to reach is to obtain a presentation of such states based on the following criteria:

- Syntheticity: permits representation of a great amount of information in an aggregate way by means of a combination of images, icons, diagrams, voice and sound.
- Analogy: allows exploitation of the human ability to comprehend pictures and sounds (resulting from a synthetic representation) and recognize specific patterns.
- Effectiveness: provides the expert with a way to easily associate patterns (above recognized) to specific configurations of a problem instance that have a meaning in the problem world.

By means of such criteria it is possible to obtain an evolutionary interaction (see Fig. 2).

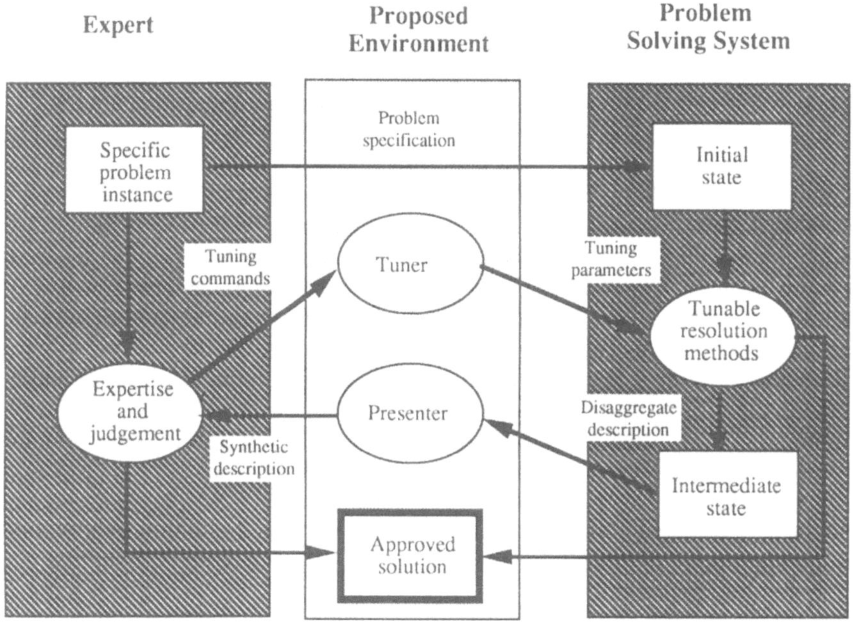

Fig. 2

The system presents (explains) a synthesis of the state it reached, and the user can interpret and judge the suitability of the method to the specific problem instance even if it was difficult to predict 'a priori' the problem-solving system's behavior. Such an interaction allows a modification of heuristics parameters as explained in [NTV89] so that the expert can tune the resolution process.

By incorporating the human in the problem-solving process [JMW89] we can improve the overall system. In fact, the expert does not have the burden of following all the steps of the problem-solving process, but only has a synthesis of the results obtained during the execution of these steps. As such a synthesis respects the analogy and effectiveness criteria, the expert can interpret these results, evaluate the heuristic adopted by the system and, when necessary, tune the heuristic according to his/her specific knowledge. Moreover, by means of such synthetic representation it is possible to represent the problem-solving skills and problem mechanisms of complex systems [CTJ87, HCP87] and as a consequence to improve the user expertise by showing the relationship between the complex system's components and the behavior of the resolution process. The problems concerning the expert system's explanation are discussed at the system level in [CTJ87], while a simple but expressive example of the visualization of optimization processes in design applications can be found in [Bur89].

3. Metaphors and Presentation Issues

In the context of high-level man-machine interfaces, metaphors are usually used to support familiar interface styles, facilitating comprehension and direct manipulation of the underlying application's concepts [CaT82].

In our framework, in order to support the presented kind of cooperation effectively, we found it necessary to provide the user with two different metaphors:

- *Problem Representation Metaphor (PRM)*

The PRM defines a way of representing and rendering to the user the entities, the relations (among entities), and the operations (ways of manipulating the entities) which make up the domain of the problem under examination; this representation is essentially concerned with the structural and static aspects of the problem domain (i.e., with the rules and constraints of the specific problem).

Examples of such a metaphor are the correspondences defined in a Macintosh between types of documents and icons, between commands and menu items, and between the folder-document relation and the directory-item relation. Other examples of this kind of metaphor are the possible ways of rendering a chess board together with its pieces: it can be represented by a list of positions (Ke1, Qd1, ..., Ke8, Qd8), it can be represented by an 8×8 matrix where letters are placed in cells corresponding to pieces' positions, or it can be represented by a synthesized 3D image.

- *Problem Perception Metaphor (PPM)*

The PPM defines a way of representing and rendering to the user the dynamics of the problem, that is, the dynamic relations existing among its entities, the evolution of such relations, and the ways in which the problem evolves; it is therefore essentially concerned with the *strategical* and dynamic aspects of the problem domain (i.e., the non-predictable evolution of the complex problem during the resolution process).

The main purpose of this metaphor is to present to the user alternative and more synthetic and powerful representations of actions and strategies and to allow him/her to absorb and master them so that: (i) simple strategy can be automated, and (ii) complex strategies may be better defined and controlled.

Examples of the Problem Perception Metaphor are: the representation of the typology of the solution of a particular problem by means of a triangle-like image (see Fig. 3 and 4), where the opening of the vertex angle represents the cardinality of the set of feasible solutions (i.e., reachable from a given initial configuration), and the kind of closure of the angle gives information on the optimality of the solutions in the set.

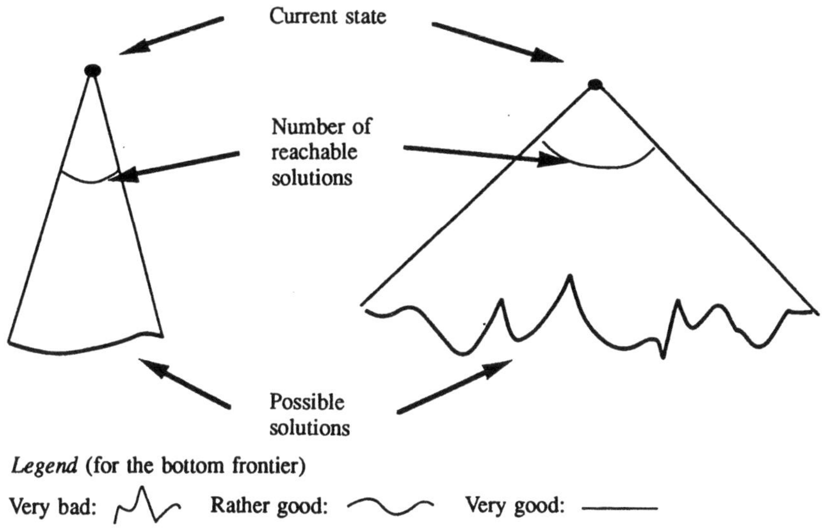

Fig. 3

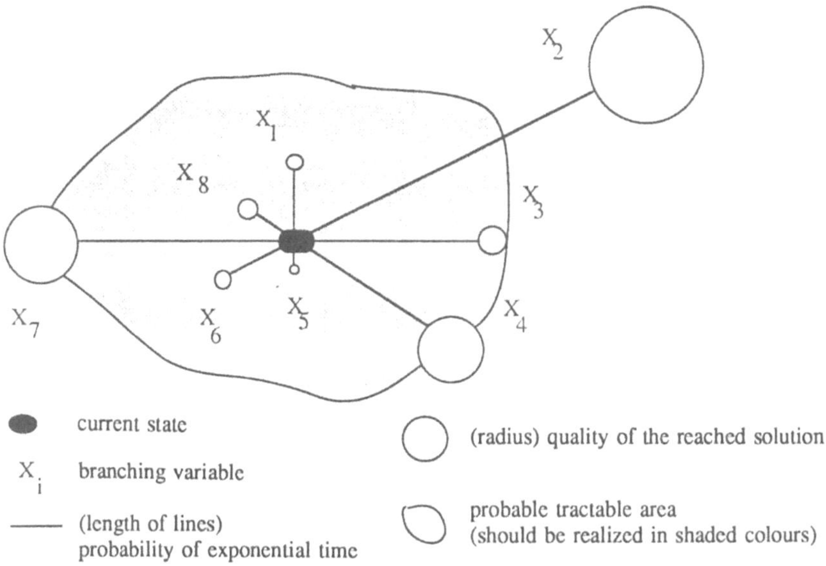

Fig. 4

Other examples, in the context of the Macintosh user interface, are the fact that an icon which has been selected highlights itself, and the fact that the trash appears fat when documents have been put into it and it has not yet been emptied.

The realization of such complex metaphors involves strong presentation requirements at the user interface level. Different studies have been done to face the problem of presentation of information for information systems [Mac86], complex systems

[HKC87, Far87] and expert systems [CTJ87, Zac85]. A multimedia approach [HSA89, BNG89, McD82] to such a problem has recently become fully exploitable by means of the support offered by both hardware of advanced scientific workstations and commercial software packages for M^4I.

Consider, for example, the APOLLO 10000 3D shading animation capabilities, the Sun Sparckstation capabilities of controlling voice and sounds, the Sun Visual Access environment or the DEC Multiwindow interface toolkit.

In order to obtain a presentation that satisfies the criteria of syntheticity, analogy and effectiveness above presented, we have to define how the information related to complex states of the problem- solving system is mapped to the user level. Such mappings can be realized by utilizing a set of interface multimedia functionalities. Different tools can be made available for this purpose, but in an integrated approach a suitable coupling of different output media channels [HSA89] must be provided in order to reach an effective breakthrough between the user and the problem-solving system.

In our environment, a number of output media couplings are provided. For instance, an overlap of raster images with vector drawings can show to the expert a (vector) partition of the complex raster image that explicitly points out all the areas where the individual components evolved. Another example is the coupling of voice and images. A voiceover speech can describe the meaning associated with components of complex diagrams. Moreover a suitable matching of sounds and images can attract the attention of the user to a partially hidden image (i.e., alert signals).

4. A Proposal for VIPS Architecture

In this section a schema of the proposed environment architecture is described. Three layers allow to define the particular functions realizing the representation of the problem and the resolution process at different levels of abstraction. Three major components, one for each level, are defined, namely Problem Solving System, Observer and Synthesizer, and Multi-media UIMS (see Fig. 5).

The *Problem Solving System*, as stated above, is seen as a module containing whatever operators and tunable heuristics are available for the resolution of specified classes of problems. It receives tuning parameters which allow the system to modify its search strategies according to user expectations. It furnishes as output a low-level description of the current state of the resolution process.

The *Observer and Synthesizer* directly interacts with the Problem Solving System, controlling the heuristic strategies and receiving unstructured data about the behavior and the state of the underlying system. This modules provides to the next higher level a synthetic and aggregate description of problem-solving activities which are meaningful at the level of the Problem Representation and Perception metaphors.

The *Multi-media UIMS* has the task of both presenting in an effective way the Problem Representation and Perception metaphors and accepting and translating the expert's commands. It receives, from the Observer and Synthesizer module, values describing the dynamic behavior of the metaphors objects and produce a multi-media presentation of the metaphors themselves. The expert interacts with the whole system,

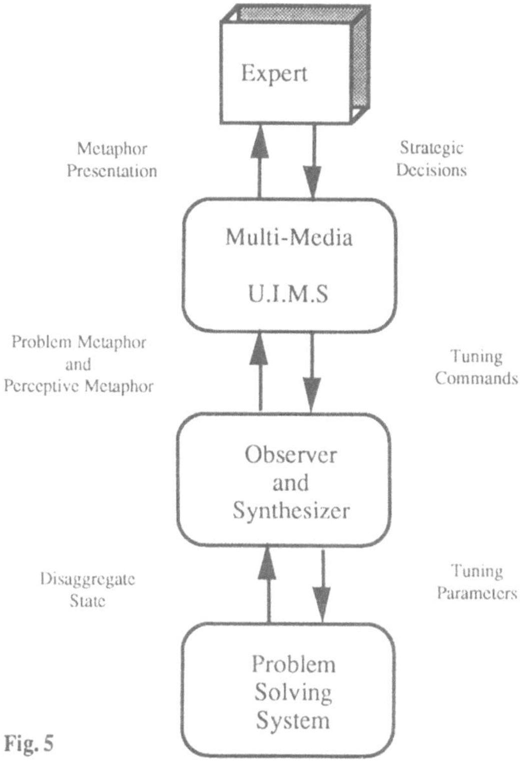

Fig. 5

providing strategic decisions, commands and queries on system behavior using multimedia interface functionalities.

In the following subsections, every layer introduced will be discussed in more detail. Further specialized submodules regarding data synthesis and aggregation, problem solving and browsing capabilities will also be described (Fig. 6).

4.1 Problem-Solving System Layer

In the architecture presented, the problem-solving system module contains a description of the problem instance and a set of tunable resolution methods. We are considering the case in which a generic resolution method can be integrated with techniques that allow the tuning of the heuristics themselves. It is quite reasonable to think that this can be performed in two different ways. We can choose which will be the node of the search graph [Nil71, Pea84] to be expanded next [NTV89, Voc89] and which operator will be used to expand it [MUB83].

The system starts from an initial state that corresponds to the problem instance and applies one of the methods available. Every intermediate state reached during the resolution process can be examined by the expert, who can evaluate the suitability of the method currently applied with the specific problem instance. Higher-level decisions of the expert are translated into sets of tuning parameters which are communicated to the problem solving-system (Fig. 6).

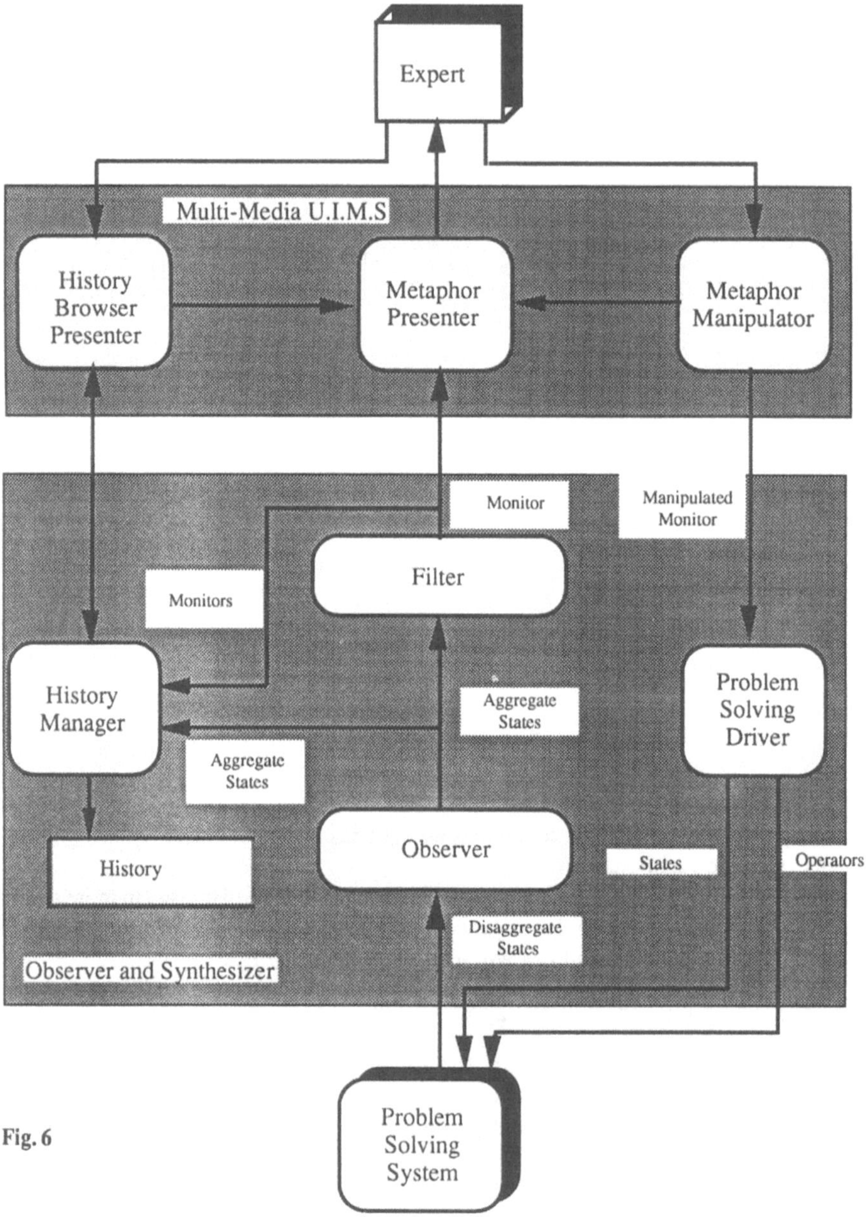

Fig. 6

4.2 Observer and Synthesizer Layer

In this layer we distinguish four modules: State Observer, Filter, History, and Problem Solving Driver (Fig. 6).

The *State Observer* has the task of structuring the low-level data from the problem-solver representation to a higher-level one, that allows a first interpretation of the process information. The Observer contains a set of predefined schemas for data representation, that discriminate and efficiently structure the information related to the process evolution. This structured representation yields, in correspondence with process evolution, the updating of different sets of data values that, for example, describe the state of the current problem instance, the operators applied, and estimators of the resolution process behavior [Voc89]. Such structured data are passed to the Filter module.

The *Filter* has the task of aggregating and synthesizing the structured data received from the State Observer. The Filter contains a set of predefined functions, Monitors, which allow the mappings of structured data into dynamic objects described by the multi-media UIMS language. The attribute values of the presentation and rendering of interface objects are related to the process evolution by means of Monitors. A key feature of Monitors is that, due to their intrinsic complexity and dynamics, they require high-level multi-media capabilities and sophisticated representation techniques.

The *Problem Solving Driver* translates high-level commands derived by the manipulations of Monitors into low-level sets of states and operators. The expert can express his/her strategic decisions by manipulating interface objects. From such manipulations it is possible to obtain states and operator values that realize the updating of the resolution method currently used by the problem-solving system according to the expert's wishes.

The *History* is a base of sequences of Monitor evolutions and aggregate states, dynamically recorded during the resolution process. Such a base can be queried and navigated in order to understand the complex behavior of the process evolution at a later time, avoiding waiting times due to the high complexity of combinatorial problems.

4.3 Multi-media UIMS Layer

In this layer we distinguish three modules: the History Browser, Metaphor Presenter, and the Metaphor Manipulator (Fig. 6).

The *History Browser* allows complex interaction between the Expert and the History base to take place. The expert can navigate in the previously recorded Monitor evolutions and aggregate states so that past experiences can be re-examined and fully understood. Different sequences of 'go-back' and 'forward' commands can be performed and the corresponding Monitor evolution and aggregate states related to recorded problem instance can be presented. The History Browser uses the presentation functionalities of the Metaphor Presenter discussed next.

The *Metaphor Presenter* permits the simultaneous presentation of Monitor evolution by means of a combination of images, voice and sounds, in a uniform, coherent and effective framework. UIMS multi-media functionalities take care of the representation of the Monitor description furnished by the Filter.

The *Metaphor Manipulator* allows the expert to directly manipulate suitable Monitor representations in order to express his/her strategic decisions and to query the problem-solving system. The values related to Monitor manipulations and queries are passed to the Problem-Solving Driver for the subsequent low-level translation.

5. Conclusions and Further Research

In this paper we have presented a framework for the design and development of an intelligent mixed-mode man-machine interface for on-line cooperation between the expert and the problem-solving system during the resolution process.

One of the main topics that requires further research is to define classes of problems for which it is possible to map the problem instance into a geometric space. By means of this mapping the configuration of the problem instance will be translated into a geometric configuration in which visual patterns can be easily recognized. The association of such patterns with expert mental representations of the problem should be investigated in the field of mental and cognitive models.

Acknowledgements

We wish to thank Marco Cini for helpful discussions about the EEC Project PONTIFEX.

Bibliography

[Bur89] S. A. Burns: Graphical representation of design optimization process. Computer Aided Design 21 1, January 1989.

[BNG89] M. L. Brown, S. L. Newsome, E. P. Glinert: An experiment into the use of auditory cues to reduce visual workload. Proc. CHI '89, May 1989

[CaT82] J. M. Carroll, J. C. Thomas: Metaphor and the cognitive representation of computing system. IEEE Trans. on Systems, Man, and Cybernetics 12, 1982

[CTJ87] B. Chandrasekaran, M. C. Tanner, J. R. Josephson: Explanation: the role of control strategies and deep models. In: J. A. Hendler (ed.) *Expert system: the user interface*. Norwood 1987

[Far87] E. J. Farrell: Visual interpretation of complex data. IBM System Journal 26 (2), 1987

[FaZ84] K. P. Fahnrich, J. Ziegler: Workstations using direct manipulation as interaction mode: aspects of the design, application, and evaluations. In: B. Shackel (ed.) Proc. First Conference on Human-Computer-Interaction INTERACT 84, London, 1984

[GaJ79] M. R. Garey, D. S. Johnson: Computer and intractability: a guide to the theory of the *NP*-completeness. San Francisco: W. H. Freeman, 1979

[HKC87] R. H. Hopkins, K. B. Campbell, N. S. Peterson: Representations of perceived relations among the properties and variables of a complex system. IEEE Trans. on Systems, Man, and Cybernetics 17 (1), 1987

[HSA89] M. E. Hodges, R. M. Sasnett, M. S. Ackerman: A construction set for multimedia applications. IEEE Software, January 1989

[JMW89] V. S. Jacob, J.C.Moore, A.B.Whinston: An analysis of human and computer decision-making capabilities. Information & Management 16, 1989

[LuS89] G. F. Luger, W. A. Stubblefield: Artificial intelligence and the design of expert system. New York: Benjamin Cummings 179–181, 1989

[Mac86] J. Mackinlay: Automating the design of graphical presentations of relational information. ACM, 1986

[McD82] N. H. McDonald: A multi-media approach to the user interface. In: Vassiliou (ed.) *Human factors and interactive computer systems*. Ablex, 1982.

[MUB83] T. M. Mitchell, P. E. Utgoff, R. Banerji: Learning by experimentation: acquiring and refining problem-solving heuristics. In: R. S. Michalski, J. G. Carbonell, T. M. Mitchell (eds.) *Machine learning: an artificial intelligence approach*. Palo Alto: Tioga, CA, 1983 and Berlin Heidelberg New York: Springer 1984, pp. 163–190

[Nil71] N. J. Nilsson: Problem solving methods in artificial intelligence. New York: McGraw-Hill, 1971

[NeS72] A. Newell, H. A. Simon: Human problem solving. Englewood Cliffs, NJ: Prentice Hall, 1972

[NTV89] E. Nardelli, M. Talamo, P. Vocca: Problem solving by interactive guide of heuristics. In: Proc. IV International Symposium on Computer and Information Sciences, Turkey, 1989

[Pea84] J. Pearl: Heuristics: intelligent search strategies for computer problem solving. Reading, MA: Addison-Wesley, 1984

[Voc89] P. Vocca: Animazione di algoritmi di visita guidata per grafi AND/OR. In: Proc. 1989 National Conference of Italian Association for Automatic Calculus (AICA), Trieste, 1989

[Zac85] W. Zachary: Beyond user friendly: designing decision aid interfaces for expert end users. In: Proc. 1985 Conference on Cybernetics and Society. New York: IEEE 1985

A GRAPH-BASED VISUAL TOOL FOR WORKSTATIONS

Naota Inamoto and Tosiyasu L. Kunii

ABSTRACT

A software tool has been designed and developed for workstation applications. It is a library of functions based on a graph formalism, and consists of graph manipulation functions and visualization functions. It provides for a standard graph editor and performs animation of changing graph data. We describe the characteristics of this tool and illustrate an extended Petri net simulator and a prototype robot system as its applications. In the prototype robot system, the graph editor is used for robot data input and the formalized graph data structure is used for data manipulation, for example in 3D set operations.

1. INTRODUCTION

Man-machine-communication is important for jobs that are done in cooperation with machines. A sound internal data model and human interface would increase the efficiency in such situations.

Roles of a workstation are classified into the following two types:

1. Development of application systems
2. Operation of application systems

For example, robot simulation is a typical workstation application. In such an application, development of a solid modeler and data input of robot shapes and motions correspond to 1 and 2, respectively. Both of these roles involve the next three factors.

1. Input and modification of programs and data
2. Debugging of programs and data
3. Cooperation with other designers (system developers)

For these purposes, we have developed a visual tool which incorporates concepts of a picture editor and visual execution. The purpose of this tool is not only visual communication but also the representation and manipulation of internal data.

The following functions are essential to a workstation.

1. Interactive graphics
2. Programming environments using conventional languages
3. Multi-window systems
4. Networks

Man-machine communication is better facilitated if AV (audio & visual) devices are available in addition to the character display. While sound synthesis and recognition are very difficult, image synthesis and drawing pictures using a mouse are comparatively easy. Readability of pictures is better than that of texts. Interactive graphics is required for visual communication.

If workstations are to be used for system development, an appropriate programming environment which incorporates conventional high-level languages is required.

Multiple windows are also of significant importance. An advantage of a multi-window system is that several texts and pictures are simultaneously visible and a variety of processes can be simultaneously executed without exiting even if the process of the current window frequently changes.

Tasks are often performed in a distributed manner and concurrently on different workstations. In such a case, a common interface amongst workstations is very important to enable cooperation in developing applications. Hence, there is a need for a network.

It is essential to combine workstations with different abilities so that their power can be utilized efficiently. For example, many systems such as robot simulation systems would require the combination of the facilities of a special-purpose parallel processing workstation, and a workstation with high speed graphics for animation. Another use of a network would be to connect such workstations.

Recent typical workstations are characterized by facilities such as the UNIX operating system [Ritc74], the C programming language [Kern78], EtherNet and the X window system [Sche86].

Even if a workstation has intelligent hardware, it cannot make the best use of its hardware if the appropriate software tools are missing. Software tools are required to perform input and debugging of data and programs effectively and efficiently. Of these, visual communication tools such as menus and graph editors are of significant importance.

A great deal of researches have been done concerning the use of visual graphs, pictures and images as methods for visual understanding [Sugi81] in user interfaces [Myer88] and visual programming [Shu88, Teit81, Glin84, Reis85, Rubi85, IEEE85]. Data structures for graphs and pictures are always necessary when such systems are implemented. Consequently, a formalization of graph data structures which incorporates graph manipulation operators will decrease the cost of application development and make application systems more conventional.

When graph editors are designed, the internal data model is more important than the specification of the editor functions. This is because editor functions can be changed according to user's requirements while the data model is fixed for portability. Well-defined general purpose graph data structures are required for the portability of graph data, which is generated by machines and edited by editors. We have performed graph data formalization and have implemented a graph editor based on this formalism. Our graph data model and editor could possibly represent a standard tool for processing and visual communication of graph data.

Design systems based on molecular aggregation such as digital and analog electronic circuits require a graph data model, a graph editor and a database. Although CAD systems based on solid modeling do not need graph editors, they also need a graph data model and a database. Formalization of the database schema for molecular aggregation has already been done [Bato84]. A database based on a graph data model is worked out by H.S. Kunii [Kuni83] and it was also implemented by software research center of RICOH Co., Ltd. [RICO88]. However, it is more efficient when data is on memory than when data is on disk. For interactive use of workstations, tools which manipulate graph data on memory are required.

The organization of the paper is as follows. Section 2 introduces the graph data structures of the tool. Section 3 explains aspects of visual communication using this tool. Section 4 shows its application to a Petri net simulator. Section 5 illustrates its application to a robot simulator.

2. A GRAPH-BASED TOOL FOR WORKSTATIONS

We believe that graph-based models are powerful candidates for constructing visual communication tools. In this section, our graph-based method is described.

2.1 Software and Hardware Design

Software and hardware systems and data descriptions implicitly include many relations such as hierarchy and groups. If comprehensibility of these relations improves, product reliability and long term productivity will increase. Most of current description methods use character-based texts, in which hierarchy can be represented by indents in texts. However, to represent relations of a graph such as many-to-many relations, it is necessary to use identifiers to relate different places in texts. It is impossible to remember many identifiers at the same time. Thus, if complexity of systems and data increase, then comprehensibility, reliability and productivity of systems and data will decrease. To maintain comprehensibility when the system and data become large, relations between different places must be clearly understood without the need to correctly memorize many identifiers. One way of achieving this is the use of graphs as pictures.

For large and complex system designs, T.L. Kunii and M. Harada formalized the design processes of complex software and hardware by a graph formalism called the **recursive graph** [Kuni80]. The term **recursive** indicates that nodes may exist within a node (a graph may exist within a node). The fundamental concepts of the graph technique presented in this paper is based on this recursive graph. Figure 2.1 shows an example of the recursive graph.

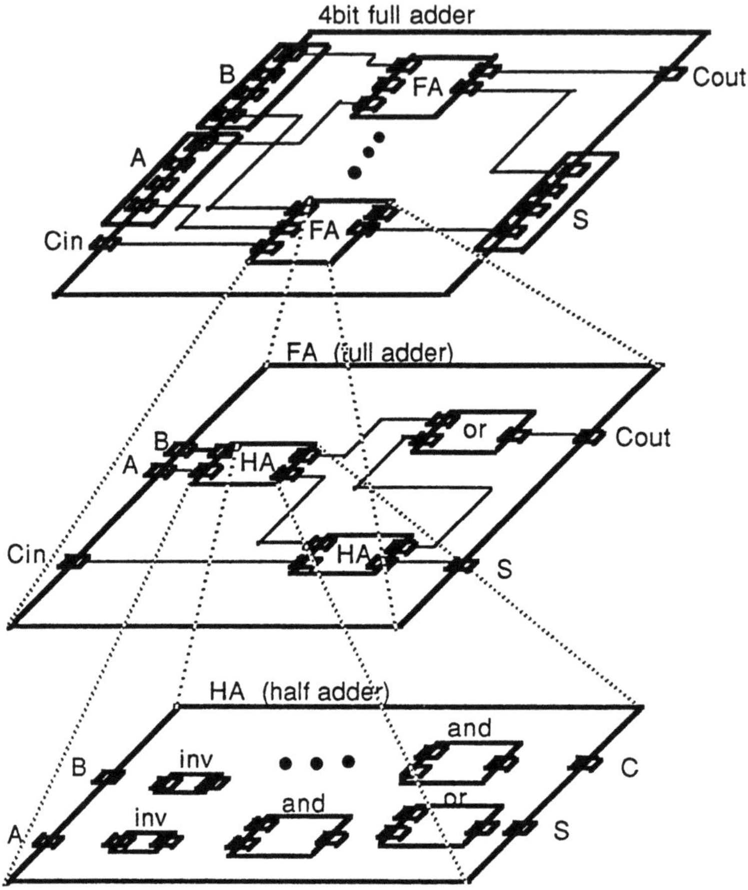

Figure 2.1 An example of the recursive graph

The original formalization of the recursive graph has the following form:

rg = <N, A, af, sn, pt, sa, ns, as>

af: $A \rightarrow N \times N$	/* arc function */
sn: $N \rightarrow 2^N$	/* subnode function */
pt: $N \rightarrow 2^N$	/* port function */
sa: $A \rightarrow 2^A$	/* subarc function */

ns: node semantics
as: arc semantics

This formalization had not been practically used for ten years, even though during that time the use of graphs for various purposes became popular. One reason why such a tool had not been implemented was that the environment consisting of mainframes and isolated graphics terminals was not suitable. But with the progress of workstations, it is possible to realize such graph-based tools today. The importance of graphs has also been indicated by other people [Nagl86].

2.2 Implementation Oriented Formalization

In this section, disadvantages of the original recursive graph formalization are enumerated from the viewpoint of implementation. A new formalization is presented, which makes implementation of the recursive graph easier.

Constraints elaborated below are essential to graph structures in most cases.

Definition

n_1->>n_2 **if and only if**

$n_2 \in sn(n_1) \cup pt(n_1)$ or

there exist $n_3 \in N$ **such that** $n_3 \in sn(n_1) \cup pt(n_1)$, n_3->>n_2

Constraints(1)

A node cannot be its own descendant.

for all $n_1, n_2 \in N; \neg(n_1 \text{->>} n_2, n_2 \text{->>} n_1)$

Constraints(2)

The number of parents of a node is lower than two.

$n_1 \neq n_2, n_1 \text{->>} n_3, n_2 \text{->>} n_4 \Rightarrow n_3 \neq n_4$

Constraints(3)

Arcs connect ports only.

for all $a \in A$; $af(a) = <p_1, p_2>$,

there exist $n_1, n_2 \in N$ **such that** $p_1 \in pt(n_1)$, $p_2 \in pt(n_2)$

Constraints(4)

The relation between parents of two connected ports is either oneself, sisters or parent-and-child.

for all $a \in A$; $af(a) = <p_1, p_2>$,

there exist $n_1, n_2, n_3 \in N$ **such that** $p_1 \in pt(n_1)$, $p_2 \in pt(n_2)$,

$$n_1=n_2 \text{ or } n_1,n_2 \in sn(n_3) \cup pt(n_3) \text{ or } n_1 \in sn(n_2) \text{ or } n_2 \in sn(n_1)$$

Constraints(5)

Ports of subarcs are subports of ports of the parent arc.

$$\textbf{for all } a_1,a_2 \in A;\ af(a_1) = \langle p_{11}, p_{12}\rangle,\ af(a_2) = \langle p_{21}, p_{22}\rangle,$$
$$(a_1 \in sa(n_2) \Rightarrow$$
$$p_{11} \in pt(p_{21}),\ p_{12} \in pt(p_{22}) \text{ or } p_{12} \in pt(p_{21}),\ p_{11} \in pt(p_{22}))$$

In the implementation based on the original recursive graph formalization, these constraints must be checked or guaranteed to maintain the correctness of graph data whenever modifications are made. Constraint-checking consumes much time and the implementation of graph modification operators which conserve the constraints is not an easy task. In order to overcome these disadvantages, a graph model is created, which incorporates the above constraints (the data in the model always satisfies the above constraints).

Horizontal relations between nodes are represented by graphs, but most vertical relations amongst nodes are trees. In the original formalization, treatment of vertical tree and horizontal graph relations are not distinguished. Thus, it is necessary to check the constraints whenever data is modified or to implement operators which conserve the constraints.

The tree structure can be defined recursively. For example, a set of n-ary trees is defined as follows:

$$NODE \equiv LEAF \cup NODE^n$$

By such recursive definitions, horizontal graph and vertical tree relations can be distinguished. The improved recursive graph is defined as follows:

$$
\begin{aligned}
RG \equiv\ & LEAF \cup \\
& \{\langle NODE, self, port, ARC, child, ns, as\rangle \mid \\
& \quad self \in NODE, \\
& \quad port: NODE \rightarrow PORT, \\
& \quad ARC \subseteq \{\langle\langle n_1, p_1\rangle,\langle n_2, p_2\rangle\rangle \mid \\
& \qquad n_1,n_2 \in NODE,\ p_1 \in pt(port(n_1)),\ p_2 \in pt(port(n_2))\}, \\
& \quad child: NODE - \{self\} \rightarrow RG, \\
& \quad ns: \text{node semantics}, \\
& \quad as: \text{arc semantics} \\
& \}
\end{aligned}
$$

$$PORT \equiv 2^{\Sigma} \times PORT \times PS$$

Σ: a set of identifiers
PS: domain of port semantics

$$pt(p_1) \equiv \{\langle\rangle\} \cup \{\langle id, p\rangle \mid \langle id, p_2, ps\rangle \in p_1, p \in pt(p_2)\}$$
/* all subports of a port */

If this formalization is used, most constraints which were described previously are automatically satisfied.

The vertical hierarchical relations between nodes represent **node inclusion**. The vertical hierarchical relations between ports represent **port inclusion**. The horizontal many-to-many relations between two ports are called **association**.

A brief explanation of this model is given below:

1. A graph has a set of nodes.
2. A node has hierarchically organized ports.
3. An arc connects two ports.
4. A node may have a child graph (ability of macro nodes).
5. A graph has a special node called "self". Arc connection with a parent graph is realized by using the "self" node.
6. Arcs can also be hierarchically organized (ability of macro arcs).

5 and 6 are related to the interfaces between modules. Although an appropriate processing system (an interpreter of graph data) is necessary, 5 and 6 are very powerful for interfacing modules.

To facilitate modular software design, it must be easy to analyze module interfaces and to arrange modules in groups flexibly. Hence, visual graph-based methods will play an important role in the future.

For practical use of graph-based tools, not only the data model but also operations for data manipulation must be provided. Graph manipulation operations such as **create_node**, **create_port** and **create_arc** are provided by the implemented prototype tool.

2.3 An Application Independent Tool

Most software tools are designed and developed by different people for each application. Hence, software has to be modified considerably as the target application changes. For example, different applications such as robotics, CAD and computer animation require very similar human interfaces. Efficiency will increase if the same interface software can be used for all these applications without altering the code. The graph-based tool also aims at providing a programming environment where it is not necessary to write the portions of the software which are application-independent.

2.4 Conceptual Graph Data Handling

The graph-based tool also aims at providing effective handling of conceptual graph data. Most data structures such as structures and lists can be represented by graphs. There are applications where data must be able to be traversed in a bi-directional manner. However, data linked by pointers generally cannot be traversed in the opposite direction. Now consider the implementation of such data structures. Existing methods consist of the following two steps:

1. Define data structures
2. Write programs or make a function library for data manipulation

In the step 2, if programmers directly write programs, they must write modules for lower level incomprehensible processing of pointers. Even if programmers make a function library, the function library must be modified whenever applications are changed.

The prototype tool developed here is based on the improved recursive graph, and this tool has the following characteristics:

1. For data structures which can be represented by graphs, it is not necessary to define a new data structure. (typeless like Lisp and Smalltalk)
2. It is not necessary to rewrite a function library for data manipulation even if applications are changed.

All manipulations can be performed by a combination of primitive graph manipulation functions such as **create_node**, **create_arc** etc.

3. VISUAL COMMUNICATION FOR WORKSTATIONS

Visualization of graph data is one way to improve understanding data. Visual graph editors are human interfaces based on 2-dimensional pictures while text editors are fundamentally based on 1-dimensional strings. By 2-dimensional animation based on graph-based pictures, human beings can visualize real-time state changes of complex systems such as parallel distributed processing systems.

3.1 Graph Editors for Data Input and Modification

Visual graphs and text files are compared focusing on the following four points.

1. Comprehension of data
2. Construction of data
3. Debugging of data
4. Modification of data

It is obvious that colorful pictures are easier to understand than flat texts with monotonous color. Graph editors spend more time than conventional text editors to construct data with same semantics. But graphs are more convenient than texts when construction of complex and large amount of data are performed parallel with the thought process. In the debugging process, finding points where errors have occurred and finding the reasons for these errors are very significant. Although efficiency of debugging depends very much on functions provided by debuggers, pictures on a graphic display will make the debugging process faster and more convenient. Data related to the modified data often must be modified similarly. Thus in the modification process, finding related data is very important. In the current methods which use text files, human beings must remember many identifiers which are search keys so that related data can be found. Using graph-based methods, related data is obtained by traversing nodes along the arcs.

The following functions are required for recursive graph editors.

Necessary window functions:
- move or resize specified windows
- change a current window
- scroll and zoom
- open a new window (create or open a graph)
- close a specified window (close a graph)

Necessary editing functions:
- create nodes, ports or arcs
- set port labels, port attributes or arc attributes
- delete nodes, ports or arcs

Convenient editing functions:
- modify position, size or orientation of nodes or ports
- modify control points of arcs
- move, resize or delete all elements in a certain region
- copy all elements in a certain region to another

Using a multi-window system which provides for 2D graphics facilities, it is comparatively easy to create a recursive graph editor which has above functions. A prototype recursive graph editor has been implemented in our laboratory using the 4.3BSD UNIX operating system and the X window system version 10 under the NEWS workstation (by SONY Co., Ltd.). The source code of the prototype tool written in the C programming language consists of about 6000 lines. Our tool can be used by many other UNIX workstations because of the high portability.

Figure 3.1 shows the software architecture of our graph-based tool. The tool consists of four layers. The lower two layers are the implementation of a recursive graph formalism. The third layer contains graph visualization functions. The top layer is the graph editor.

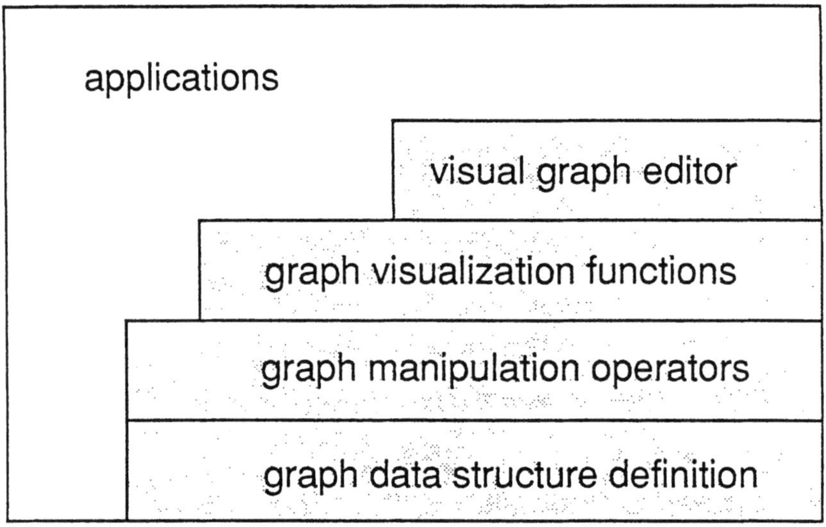

Figure 3.1 Software architecture of the graph-based tool

3.2 A Software Function Specification Method

In this section, graphical specification of data transformation functions, which is one of graph editor applications, is illustrated by an example. A link function for doubly linked lists is shown. Typical description in a manual is shown in (1). Typical source code in a program is shown in (2).

(1) Description in a manual written in natural language:

A function **link(e, l)** inserts the new element **e** in the doubly linked list **l**.

(2) Program source code:

```
link(e, l) {
  e->prev = l->prev;
  e->next = l;
  l->prev->next = e;
  l->prev = e;
}
```

Description (1) includes ambiguity. Both description (1) and source code (2) are difficult to understand quickly. In description (1), the reader is assumed to be acquainted with the term used and source code (2) is close to **von-Neuman machines**. The gap between (1) and (2) is very big. Graph-based visual specification corresponding to the description (1) and the source code (2) is shown in Figure 3.2.

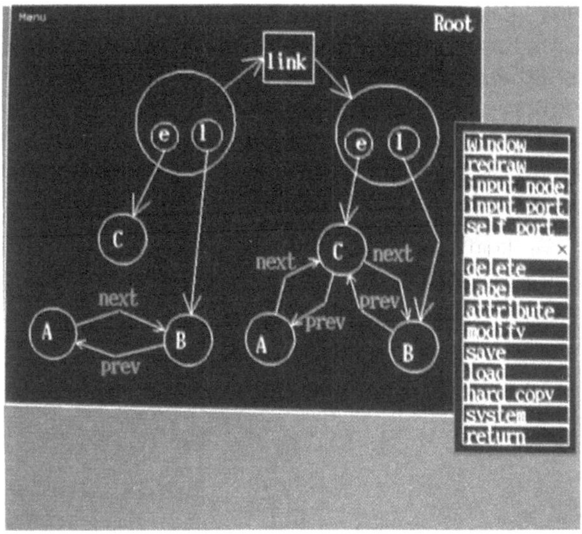

Figure 3.2 A graph-based function specification

Graph-based specification is not ambiguous and it is possible to understand the semantics of the function quickly. Actually programmers often draw and use pictures equivalent to this when they write source code such as (2) from specifications such as (1). A graph-based formalism described in this paper ensures the next step which consists of automatic source code generation from graph-based specification.

Specification description methods based on logical and mathematical formula can also be produced. These methods can specify the output only if all input data does not change until the output is completely produced, but can not specify local modifications of large amount of data without replacing the entire data. Most of operations on linked data are in the form of local modifications on large data. For this reason, the gap between logical mathematical specification and source code is very large while the gap between visual graph-based specifications and source code is small.

4. APPLICATION TO A PETRI NET SIMULATOR

4.1 Animation Techniques

It is important for the users to monitor and trace state changes of complex parallel distributed systems. Printing texts is not suitable for showing parallel changes in state. Computer animation will be the best way to show parallel dynamic information to users. As an example, an extended Petri net model is shown below. Figure 4.1 represents an example of an extended Petri net model [Yosh89] where tokens have colors (data) and arcs between places and transitions have functions. This model is very similar to a data flow model. Using 2D animation, users can grasp and monitor the non-deterministic state changes of the extended Petri net.

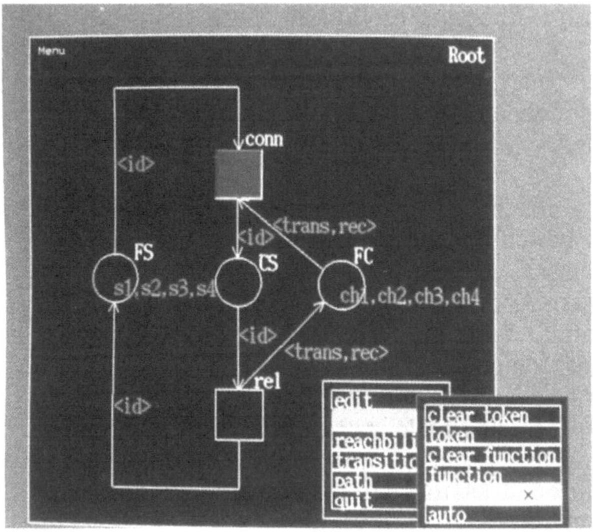

(a) Transitions which are able to fire

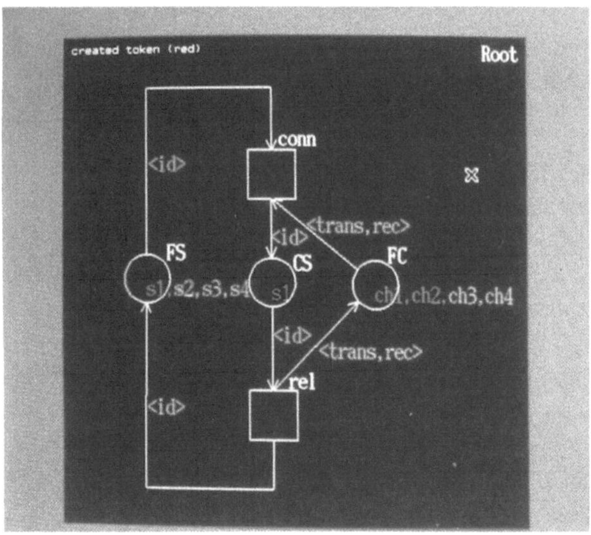

(b) Destroyed and created tokens

Figure 4.1 Animation of an extended Petri net

4.2 Transformation and Analysis of Graph Data

It is possible to execute (interpret) and transform graph data and to carry out its animation, although appropriate application-dependent parts of programs must be created.

Figure 4.2 represents a reachability tree created from the Petri net in Figure 4.1. The states which appeared in the past are all stored in the memory. If the state identical with the stored state appears, it is grouped by a different kind of arcs. And states reachable from this state are not furthermore investigated. Thus this reachability tree is essentially equivalent to a state transition graph.

Although the displayed output of the reachability tree is automatically generated by a machine, the generated graph can be edited to improve readability. The execution, transformation and analysis phases of this application are as follows:

Execution:
Animation of marking changes of the extended Petri net

Transformation:
Automatic generation of reachability trees (state transition graphs)

Analysis:
Termination check, deadlock detection, and search of
the state transition paths which lead to the specified marking etc.

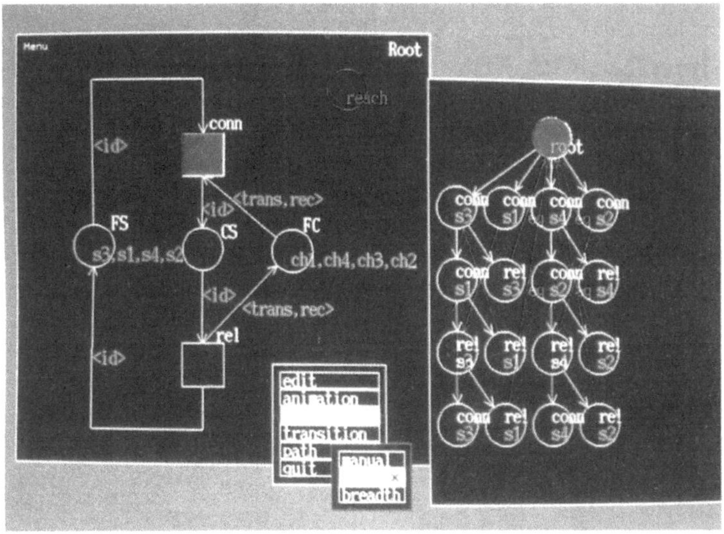

(a) A reachability tree by depth first search

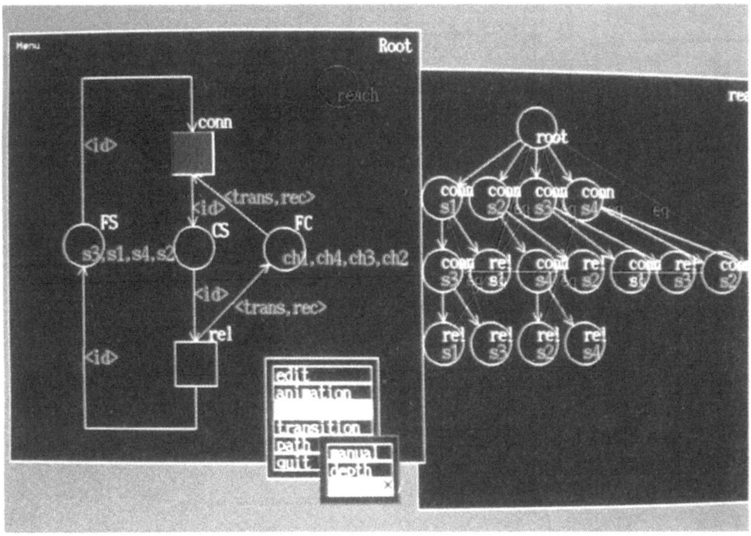

(b) A reachability tree by breadth first search

Figure 4.2 Reachability trees of an extended Petri net

5. APPLICATION TO A ROBOT SIMULATOR

5.1 A Human Interface with Visual Graphs and Images

Text-based programming languages with standard graphics libraries currently must be used for computer animation of robots. A robot programming method in operation in our laboratory uses as many graph-based pictures as possible instead of classical text files. Four kinds of graphs are currently used. They are illustrated in Figure 5.1. It is very important that these graphs are visible and can be modified simultaneously using a **multi-window system.**

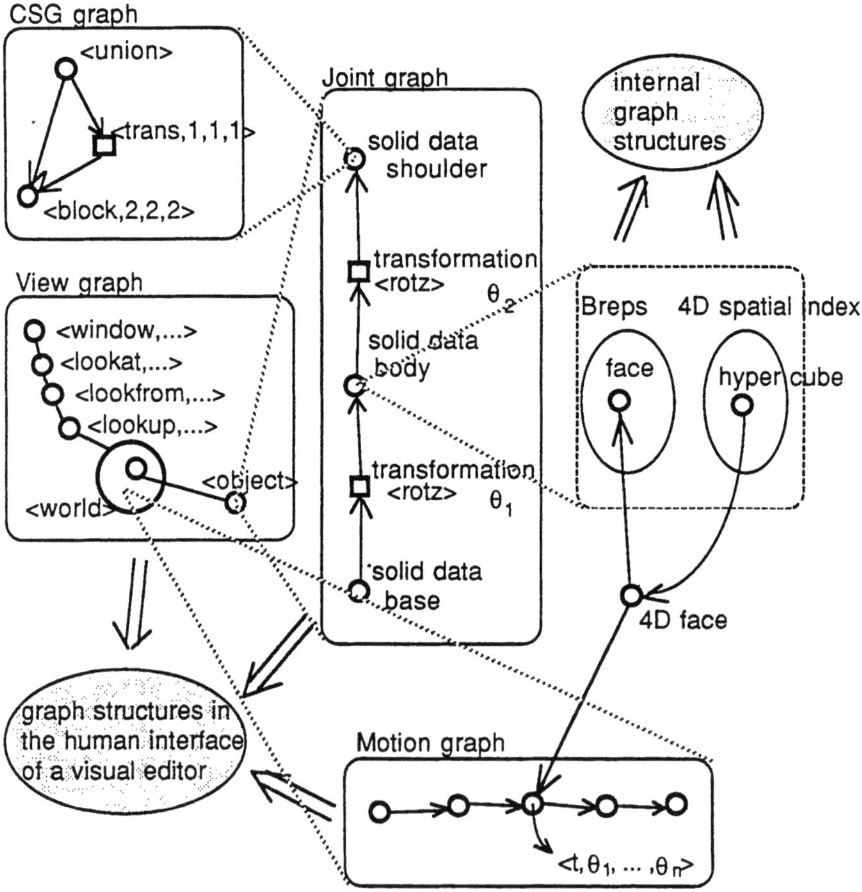

Figure 5.1 Visual graphs for data input and debugging

The first one is called a **view graph**. View graphs specify how to project the world onto the screen. Using a view graph, multiple projections can be specified easily by connecting window nodes and a world node, and interactive (real-time) changes of viewing parameters such as target position and eye position are possible. Grouping multiple objects is also easy by connecting a world node and object nodes.

The second one is called a **joint graph**. Joint graphs specify how links of robots are connected. The third one is called a **CSG graph**. CSG graphs define the shapes of the links. One kind of node in a joint graph represents a link of a robot. There is a CSG graph in this node. During kinematic and dynamic robot simulation in robotics, robot types and shapes are frequently changed. For interactive design and test, the results of simulation should be obtained in-real-time. If calculation speed of a robot simulator is very fast, human interfaces will become the bottle neck. Joint and CSG graphs are considered to improve data input and modification in comparison with texts and to facilitate easy understanding. For example, the shape definitions of robot links are easily obtained from the corresponding node in a joint graph.

The fourth one is called a **motion graph**. Motion graphs describe multiple robot motions at a point-to-point level. Because of the improved recursive graph, it is easy to specify interfacing between several robots and their motions. It is also easy to specify synchronizations among multiple robots by arcs representing before-and-after relation between two nodes of two motion paths. Motion graphs can be regarded to be a kind of control graphs. By marking firing nodes in a motion graph during animation and interference check, it is easy to know the current point of execution in a motion graph and to debug robot motions interactively by the graph editor when errors (collision with obstacles etc.) occur.

While texts require compliers, a graph editor does not require them. Graph data is processed by an appropriate interpreter with few transformation, and the interpreter makes the system interactive.

Screens of the implemented robot system is shown in Figure 5.2.

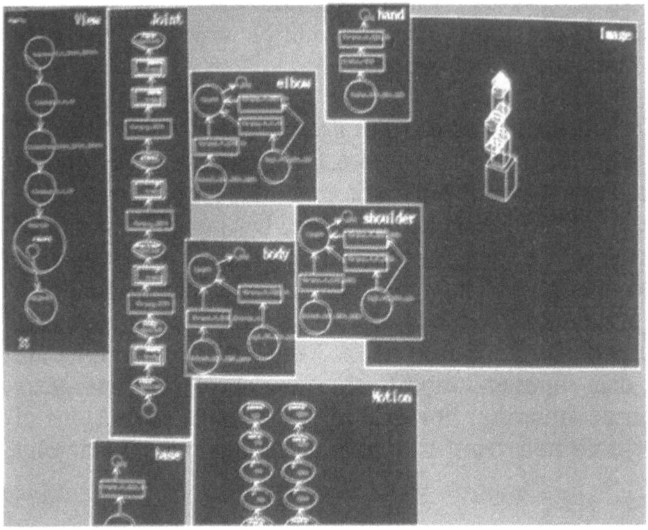

(a) A robot defined by joint and CSG graphs

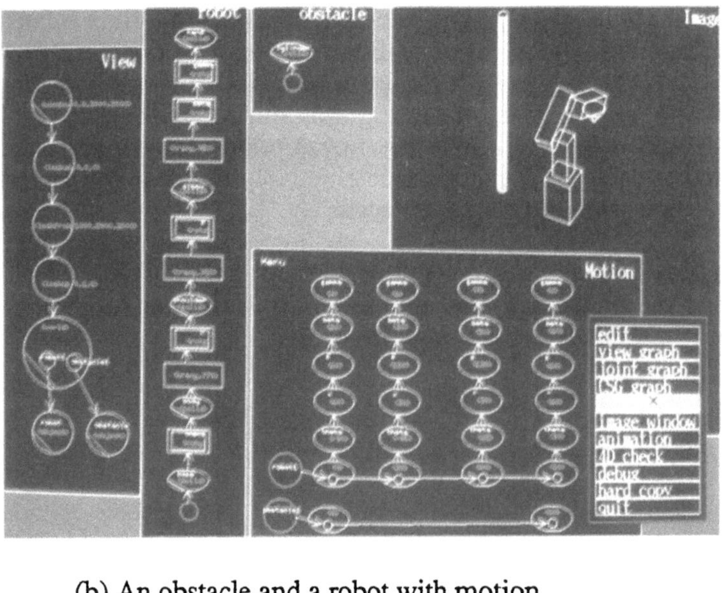

(b) An obstacle and a robot with motion

Figure 5.2 An example of four kinds of graphs for robot programming

5.2 Internal Data Structures of a 3D Solid Model

We have implemented prototype 3D solid modeler as the first step to the implementation of a 4D solid modeler which includes time. This modeler uses boundary representation (B-reps) [Brai80,Mant82] as a basis of shape representation and uses octree [Jack80,Meag82,Yama84,Fuji85] as a hierarchical spatial index. Figure 5.3 illustrates the internal data structure of a solid model based on the improved recursive graph.

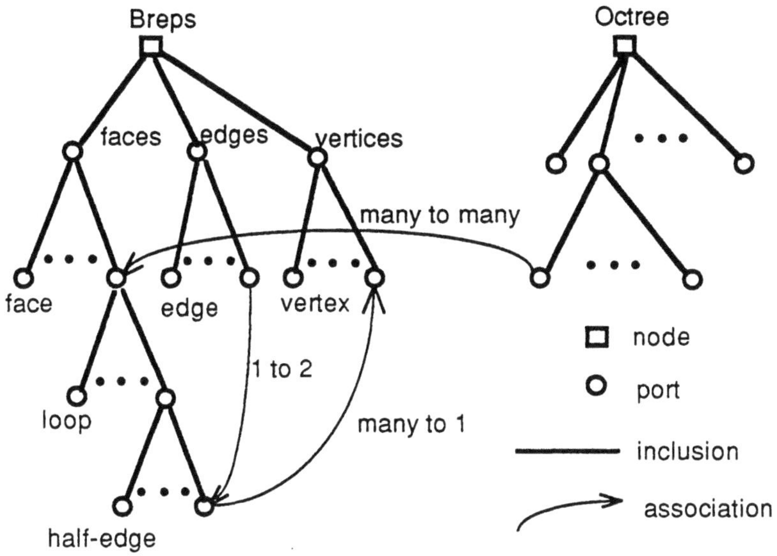

Figure 5.3 Internal data structures of a 3D solid model

Combination of boundary representation (B-reps) and octree is difficult and very few implementations exist. With the graph-based tool, we can use the data structure of this tool as it is, for internal representation of the 3D solid model. Hence, B-reps and octree systems developed independently can be combined without modification. Our method of integrating B-reps and octree connects faces and cubes which intersect each other. Although connections are many-to-many and required to be traversed in a bi-directional manner, graph data manipulation is very easy using the graph-based tool. Simple example algorithms using this integration are shown in Figures 5.4 and 5.5.

```
set_operation(oct1, oct2 )
{
    if (oct1, oct2 are homogeneous) {
        for all association <oct1 ,face1 >
        for all association <oct2 ,face2 >
            face_face_process(face1,face2 );
        end for
        end for
    }

    divide_octree(oct1 );
    divide_octree(oct2 );

    for all i in [0,7]
        set_operation(oct1 [i ], oct2 [i ]);
    end for
}
```

Figure 5.4 Set operations of B-reps with octree

```
divide_octree(oct )
{
    if (oct is heterogeneous) return;

    create eight child octree of oct ;

    for all association <oct , face >
        delete arc <oct ,face >;
        for all i in [0,7]
            if (face intersect oct [i ])
                create arc <oct [i ],face >;
        end for
    end for
}
```

Figure 5.5 Octree division with B-reps

Polytree [Carl85,Carl87,Durs89] is another approach to integrate the B-reps and the octree. A significant difference between the polytree and our model is that the following constraints are usually included by the polytree definition but our model is free from these constraints.

1. Octree cubes must not include more than one vertex.
2. Octree cubes with no vertices must not include more than one edge.
3. Octree cubes with no vertices and edges
   must not include more than one face.

Polytree often divides 3D space too finely due to these constraints and hence, tends to be large. Our method can allow the user to choice the conditions that determine the termination of the division process. In addition, it allows cubes to contain many vertices, edges and faces.

Our 3D model can give the range of minimum distance among objects as a result of 3D interference check. It is extended to 4D and can give a range of minimum distance among moving objects as a result of 4D interference check. This characteristics is suitable for path planning where safety levels can be selected. The term "safety levels" denotes minimum distance among moving objects here.

5.3 Visual Assembly Planning

This section describes another application of the graph editor. Robot planning is usually identical with problem solving. For example, the system STRIPS [Fike71] finds a path (a sequence of selections) as optimal as possible between the start and goal states. Most conventional ways of problem solving are searching and-or-tree with back tracking mechanism. A prolog program such as shown in Figure 5.6 represents a simple example of a robot planning. This program is not completed and only includes propositions correspondent to add-list of STRIPS. Although prolog shows the result of refuting as well as the values of variables, prolog is not suitable for treating an and-or-tree itself as data. Figure 5.6 is not comprehensive.

```
hold(s1,x) ← ontable(s0,x),clear(s0,x),handempty(s0), pickup(s0,x,s1).

ontable(s1,x) ← hold(s0,x), putdown(s0,x,s1).
clear(s1,x)   ← hold(s0,x), putdown(s0,x,s1).
handempty(s1) ← hold(s0,x), putdown(s0,x,s1).

handempty(s1) ← hold(s0,x),clear(s0,y), stack(s0,x,y,s1).
on(s1,x,y)    ← hold(s0,x),clear(s0,y), stack(s0,x,y,s1).
clear(s1,x)   ← hold(s0,x),clear(s0,y), stack(s0,x,y,s1).

hold(s1,x) ← handempty(s0),on(s0,x,y),clear(s0,x), unstack(s0,x,y,s1).
clear(s1,y) ← handempty(s0),on(s0,x,y),clear(s0,x), unstack(s0,x,y,s1).

on(s0,A,B),on(s1,B,A).
```

Figure 5.6 Simple assembly planning in prolog

Figure 5.7 shows graphical programs whose semantics are the same as that of Figure 5.6 and its interactive visual execution is possible by subgraph pattern matching and subgraph transformation (replacement) rules. A graphical program shown in Figure 5.7 presents states and actions equivalent to Figure 5.6 but execution control rules are not given here.

In graphs of Figure 5.7, circle nodes are considered as nouns, and rectangle nodes are considered to represent verbs and adjectives, and directions of arcs indicate the directions of qualifiers, and most arcs are considered to be prepositions and auxiliary verbs.

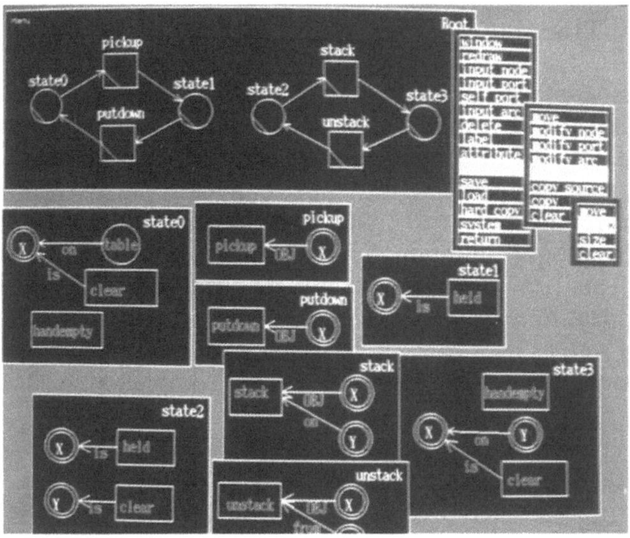

Figure 5.7 Assembly planning with a visual graph

Control and data flow can be easily understood using animation technique. This is useful not only for debugging but for intuitively understanding the information flow. Visual execution like this can be also applied to simulations of parallel distributed systems such as both von-Neuman-type multiprocessors and non-von-Neuman oriented architecture.

6. CONCLUDING REMARKS

Characteristics of the graph-based tool are summarized as follows:

1. A visual graph editor of recursive graph data for interactive design and data input
2. Transformation of graph data and its animation
3. An application-independent tool
4. Conceptual graph data handling for prototyping

Characteristics 1 and 2 were shown by examples of a Petri net simulator and a visual specification of a robot planning program. Characteristics 3 allowed this tool to be applied in cases where the semantics of the application were represented by graphs. Due to 4, coding of prototype 3D and 4D solid modelers become easy. Source code was about 3000 lines excluding the graph-based tool. These features allow easy construction of human interfaces and the kernel of a prototype robot simulation system.

This robot simulation system included 1) visual data input (views, shapes, motions etc.), 2) 3D set operations, 3) animation of robot motions, 4) 4D interference check. The graph-based tool was used as the basis of this prototype robot simulation system.

We proposed a data flow language as a conceptual model [Yama85]. This was an application-independent and device-independent model and was applied to the multiple robots control. Our graph-based tool provided facilities such as internal data structures and human interfaces, which are necessary to realize this conceptual model.

ACKNOWLEDGEMENT

We would like to thank Dr. Hiroyuki Yoshida, Ms. Deepa Krishnan, Mr. Satoshi Asami, Mr. Martin J. Durst, Mr. Tetsuhiko Ichikawa and Mr. Hitoshi Saji for their comments and discussions with them. Our thanks are also due to Dr. Hideko S. Kunii, the director of the Software Research Center of RICOH for her support to this reserch.

References

Bato84.
D.S. Batory and A.P. Buchmann, "Molecular Objects, Abstract Data Types, and Data Models: A Framework," *Proc. of the 10-th Int. Conf. on VLDB*, pp. 172-184, August 1984.

Brai80.
I.C. Braid, R.C. Hillyard, and I.A. Stroud, "Stepwise Construction of Polyhedra in Geometric Modelling," in *Mathematical Methods in Computer Graphics and Design*, ed. K.E. Brodlie, 1980.

Carl87.
I. Carlbom, "An Algorithm for Geometric Set Operations Using Cellular Subdivision Techniques," *IEEE CG & A* , vol. 7, no. 5, pp. 44-55, May 1987.

Carl85.
I. Carlbom, I. Chakravarty, and D. Vanderschel, "A Hierarchical Data Structure for Representing the Spatial Decomposition of 3-D Objects," *IEEE CG & A* , vol.5, no. 4, pp. 24-31, April 1985.

Durs89.
M.J. Durst and T.L. Kunii, "Integrated Polytrees: A Generalized Model for the Integration of Spatial Decomposition and Boundary Representation," in *Theory and Practice of Geometric Modelling*, ed. H.P. Seidel, Springer-Verlag, Heiderberg, 1989.

Fike71.
R.E. Fikes and N.J. Nilsson, "STRIPS : A New Approach to the Application of Theorem Proving to Problem Solving," *Artificial Intelligence* , vol. 2, no. 3-4, pp.189-208, Winter 1971.

Fuji85.
K. Fujimura and T.L. Kunii, "A Hierarchical Space Indexing Methods," in *Computer Graphics (Proc. of Computer Graphics Tokyo '85)* , ed. T.L. Kunii, Springer-Verlag Tokyo, 1985.

Glin84.
E.P. Glinert and S.L. Tanimoto, "Pict: An Interactive Graphical Programming Environment," *IEEE Comp.*, vol.17, no. 11, pp. 7-25, 1984.

IEEE85.
IEEE, "Visual Programming," *IEEE Comp.*, vol. 18, no. 8, August 1985.

Jack80.
C.L. Jacks and S.L. Tanimoto, "Oct-Trees and Their Use in Representing Three-Dimensional Objects," *Computer Graphics and Image Processing* , vol. 14, pp. 249-270, 1980.

Kern78.
B.W. Kernighan and D.M. Ritchie, in *The C Programming Language* , Prentice-Hall, 1978.

Kuni80.
T.L. Kunii and M. Harada, "SID: A System for Interactive Design," *NCC, 1980* , pp. 33-40, 1980.

Kuni83.
H. Kunii, "Graph Data Language: A High Level Access-Path oriented Language," Ph.D. Dissertation, Department of Computer Science, the University of Texas at Austin, May 1983; also to be available as a book from Springer Verlag.

Mant82.
M. Mantyla and R. Sulonen, "GWB - A Solid Modeler with Euler Operators," *IEEE CG & A* , vol. 2, no. 7, pp. 17-31, Sept. 1982.

Meag82.
D. Meagher, "Geometric Modeling Using Octree Encoding," *Computer Graphics and image Processing* , vol. 19, pp.129-147, 1982.

Myer88.
B.A. Myers, in *Creating User Interfaces by Demonstration* , Academic Press, INC., 1988.

Nagl86.
M. Nagl, "A Software Development Environment based on Graph Technology," in *Graph-Grammers and their Application to Computer Science (Lecture Notes in Computer Science 291)* , ed. H. Ehrig et al., pp. 458-478, Springer-Verlag, 1986.

RICO88.
RICOH, in *G-BASE Manual's* , ed. RICOH Co., Ltd., April 1988.

Reis85.
S.P. Reiss, "PECAN: Program Development Systems That Support Multiple Views," *IEEE Trans. on Softw. Eng.* , vol. 11, no. 3, pp. 276-285, 1985.

Ritc74.
D.M. Ritchie and K. Thompson, "The Unix Time-Sharing System," *CACM* , vol. 17, no. 7, pp. 365-375, July 1974.

Rubi85.
R.V. Rubin, E.J. Golin, and S.P. Reiss, "ThinkPad: A Graphical System for Programming by Demonstration," *IEEE Software* , pp. 73-78, 1985.

Sche86.
R.W. Scheifler and J. Gettrys, "The X window system," *ACM Trans. on Graphics* , no. 2, April 1986.

Shu88.
N.C. Shu, in *Visual Programming* , Van Nostrand Reinhold Company, New York, 1988.

Sugi81.
K. Sugiyama, S. Tagawa, and M. Toda, "Methods for visual understanding of hierarchical system structures," *IEEE Trans. on Systems, Man, and Cybernetics* , vol. SMC-11, no. 109-125, Feb. 1981.

Teit81.
W. Teitelman and L. Masinter, "The Interlisp Programming Environment," *IEEE Comp.*, vol. 14, no. 4, pp. 25-34, 1981.

Yama85.
K. Yamaguchi, N. Inamoto, and T.L. Kunii, "A Data Flow Language for Controlling Multiple Interactive Devices," *IEEE CG&A* , vol. 5, no. 3, pp. 48-60, Mar. 1985.

Yama84.
K. Yamaguchi, T.L. Kunii, K. Fujimura, and H. Toriya, "Octree-Related Data Structures and Algorithms," *IEEE CG & A* , vol. 4, no. 1, pp. 53-59, Jan. 1984.

Yosh89.
H. Yoshida, *Protocol Specification Methodology based on the Extended Recursive Graph Formalism* , Ph.D. Dissertation, Department of Imformation Science, the University of Tokyo, 1989.

Temporal Issues of Animate Response

A.A.M. Kuijk

Abstract

Increased capacities of personal workstations enable graphical user interfaces to offer natural human computer interaction. This leads to the possibility of animate response, i.e., natural transitions from one state to another. It is recognized that for optimal efficiency such animate response should be "tuned" to the time frame of the user. It is not sufficient to speed up the response as much as possible: the temporal characteristics of the response should be based on human perceptual capabilities. This paper is a survey of notions related to animate response. Throughout the paper we will touch upon subjects which need consideration and/or further investigation.

CR Categories and Subject Descriptors:

I.3.3 [Computer Graphics]: Picture/Image Generation – Display algorithms;
I.3.5 [Computer Graphics]: Computational Geometry and Object Modelling – Curve, surface, solid, and object representations;
I.3.6 [Computer Graphics]: Methodology and Techniques – Interaction techniques

Key Words and Phrases:
continuous motion, temporal anti-aliasing, implicit animation, direct manipulation, hierarchical data structures, progressive refinement.

1 Introduction

User interfaces are becoming more and more sophisticated due to the tempestuous evolution of both hardware and software. Sophisticated user interfaces support natural interaction and take into account features of human visual perception. This results in animate response, that is response where changes as invoked by the user are visualized by a smooth, natural transition from one state into another.

Motion plays an important role in numerous aspects of human vision. One's eyes are in a constant state of movement to generate stimuli needed to see anything at all. Relative motion of objects is essential for three-dimensional perception: a change of viewing position generates motion parallax. The human mind is trained to interpret this motion parallax as a measure of distance. A person observing a three-dimensional object which he holds in his hand will almost automatically rotate it to generate relative motion and thereby get the best impression of the three-dimensional structure. The human vision system uses relative motion as a basis of grouping–organizing objects that belong together. By exploiting such acquired perceptual capabilities of the viewer, animate response can improve human computer interaction.

Why should we be concerned about temporal aspects of animate response–isn't it sufficient to simply let the system respond as fast as it can? If current workstations are not fast enough, future systems will be, so why bother about it? The answer to this is that systems might never be fast enough. Workstations may increase in capacity, but on the other hand applications become increasingly more complex. More importantly, however, to let an interactive environment respond as fast as it can will almost never result in an adequate response. The system will quite often be too slow, but occasionally it may even be too fast.

We should continue to be concerned about temporal aspects of animate response. The system must be "tuned" to the time frame of the user. This tuning should aim at optimal interaction, on the one hand taking into account the limited temporal resolution of the display system and the human vision system and on the other hand taking into account ergonomic aspects to help the user anticipate to a new situation. The following two examples of such tuning may be illustrative. They can be found in a type of user interface commonly known as the desktop metaphor, a simple two dimensional environment.

Not all systems are capable of supporting real-time dragging of a window with all of its content. An elegant solution is to leave the window with its content where it is and let the user drag a simplified representation of the window, its boundary rectangle. Due to this simplification, the system is able to give appropriate feedback to the user.

The visible action upon closing a window, i.e., clearing the screen, can be done very quickly. Here, a system can take more time and smoothly lead the user to the new situation by an implicit animation–a shrinking boundary rectangle moving towards a disk or folder icon. During this animation, both the system and the user have

time to update their administration.
For animate response in three-dimensional dynamic environments more sophisticated techniques are needed, but basically they are of the same nature: adapt to the user's time frame. In this paper we will address issues related to this.
The outline of the paper is as follows. First we will be more specific on what we consider to be animate response. Next, we will visit an area in computer graphics which is closely related, the area of computer animation. Computer animation is an area where the knowledge of how to adapt to the human vision system is essential. Following this we will discuss aspects of visualization of motion given the limitations of a real-time environment. From this we will see that hierarchical data structures are needed so as to be able to trade time against quality. Therefore several aspects of hierarchical data structures are discussed. The last section will mention some issues which need further investigation.

2 Animate Response

As mentioned in the introduction, animate response is the visual response of a system to changes invoked by the user. It is characterized by a smooth, natural transition from one state to another. It is useful to distinguish between the following two situations: implicit animation, where the user interface automatically generates intermediate states and direct manipulation, where the user generates intermediate states by manipulating objects or entities of objects. These two categories differ in who is in control over the transition phase, a difference which has direct implications for the temporal aspects of the interaction.

2.1 Implicit Animation

In the case of implicit animation, explicit specification of a new state implicitly specifies the transition. Although the user interface may allow the user to specify certain transition parameters, the transition itself is autonomously generated by the system. As the visualization is not part of a time restricted feedback loop, the temporal requirements are somewhat liberal.
Implicit animation is a sophisticated type of response which can be employed by traditional command language interaction: objects are manipulated by means of explicit commands, issued for example by typing or by menu selection. Implicit animation merely helps to "guide the eye", to make the transition appear more natural. It reduces the computational effort of the interaction process, not on the side of the system but on the side of the user. For this, some intelligence on the system side of the interface is required. Based on the type and amount of the change, the system should be able to determine how to best visualize the transition for the user.
Implicit animation does not have to be restricted to generating a natural transition between two "key-frames". One may also think of more abstract transitions as in the

example of closing a window as mentioned in the introduction. In the first state the user sees a window, in the final state the user sees an icon and the abstract transition is used to inform the user which icon relates to the window. A characteristic of abstract transitions is that an abstract relation between two intrinsically different entities can be visualized.

As an example of non-abstract implicit animation, think of a user looking at an image of a complex molecule. To help him perceive the location of all the atoms of that molecule, the command "rotate 90 degrees around the z-axes" does not result in a mere presentation of the new situation, but in a simulation type of response: an animated sequence of a slowly rotating molecule. Even if this rotation is visualized using a simple wireframe image, the relative motion of the individual atoms during this rotation would give the viewer the depth perception needed to fully comprehend the three-dimensional structure. It may be clear that this simulation type of response needs temporal tuning. It should not take too much time else the user will be annoyed. On the other hand it should not be too fast else the user will miss essential information.

As mentioned before, setting of transition parameters may allow the user to specify the speed, image quality or other aspects of the transition. Sophisticated systems would automatically adapt to the user's skill and preferences. It should be noted that although the temporal requirements may be somewhat liberal, lengthy transition times would distract rather than guide the user.

2.2 Direct Manipulation

Direct manipulation is a very natural and powerful interaction mechanism. The user is able to manipulate a visible representation of a virtual model as if it is a real object which can be seen and touched. The user can manipulate entities of that model by means of input devices like a mouse, a light pen, a trackball or more exotic devices such as a data-glove. These manipulations have to be visualized as a part of the feedback loop between input device, visualized entity, user and again input device. The user will handle the input device based on what he sees. As a result the time restrictions are severe.

Direct manipulation has already proven to be successful in a well known type of user interface, the desktop metaphor. This type of user interface became such a successful entrant in the office environment mainly because a virtual two dimensional environment was created which was a metaphor of an environment already familiar to the user–his desk. It allows a user to manipulate documents (windows) on a screen just as he is used to manipulating paper documents on his own desk: file them in a cabinet, throw them in a waste paper basket, open and close documents, etc. This familiar "look and feel" reduces the learning time, an important consequence in the office environment.

Similarly this direct manipulation concept can be extended to three-dimensions to result in a very natural and intuitive user interaction mechanism within three-

dimensional and even dynamic environments. It is the key to creating artificial realities (also known as virtual environments or cyberspaces) in which the user is free to manipulate simulated three-dimensional objects in a natural way, can walk through and interact with imaginary worlds–in short, can be a part of the virtual model.
Based on the direct manipulation concept we can create remote control systems with which a user is able to control remotely in space (hazardous or far away environments), remotely in time (future) or allow control in a different scale (macroworld or microworld) [Shneiderman 1989].
In order to give the user the feeling that he is part of a virtual model and interacts with it, the most important aspect of direct manipulation appears to be direct visualization. If the result of the manipulation is not visualized immediately, the user will lose contact with the environment. As Scott Fisher states [Fisher 1989], "The quality of the graphics appears not to be extremely important, as long as the response is good".

2.3 Summary

In the area of animate response we distinguish two classes: implicit animation and direct manipulation. In the class of implicit animation, the system is in control of the transition phase. As a result, intermediate states are known beforehand and in principle the temporal behavior is predictable. In the class of direct manipulation however, the transition is controlled by the user. As a result the intermediate states are unknown and the temporal behavior is unpredictable. However, the impact of temporal requirements can be generalized to animate response as a whole.

3 What Can We Learn From Computer Animation?

Why is a film, a long band of static images, often called a "movie"? Answering that question also gives an answer to why computer animation is at all possible, using video displays which can only display static images. Apparently it is possible to give a viewer an impression of continuous movement by successively presenting displaced static images (frames). Given this, at what rate should these frames be generated? Intuitively it is clear that the frame rate has to be related to the speed at which the objects move. Since video displays have a refresh rate of about 100 Hz at maximum, we find ourselves restricted in frame rate [1] as well.

This limitation effects the visualization of fast moving objects. Due to their rapid motion, the displacement of objects between two individual frames becomes quite

[1] Note that frame rate is the rate at which successive and thus different frames are displayed. This is not to be confused with the refresh rate, which is dictated by the persistence time of the display medium. The frame rate can never exceed the refresh rate, but one frame may last for several refresh cycles.

large. When these object are displayed just as if they were static, the perception of the spatial relation between images is lost–the impression of watching objects in continuous motion is no longer there. A movie projector, however, shows separate frames at a rate as low as 24 frames per second. Even this rate is sufficient to generate an impression of continuous motion. Why this is so can been found by looking at individual images of a fast moving object on a camera film. There we see that the finite exposure time of the film results in blurred images of the fast moving object. Due to this motion blur the impression of continuous movement is maintained, even at a rate of 24 frames per second. From this it is concluded that the human vision system can trade off temporal resolution against spatial resolution.
A number of papers present methods to generate motion blur (also known as temporal anti-aliasing) [Korein 1983; Potmesil 1983; Max 1985; Grant 1985], but arguments why it is needed do not go any further than "animation which simulates motion blur feels more natural" or "it smooths out jerkiness". A more theoretical foundation of why and to what extent motion blur is needed can be found in E.H. Blake's Ph.D. thesis on computing adaptive detail [Blake 1989]. The next subsection is a short summary of relevant issues from this thesis.

4 A Viewer-Centered Metric for Computing Adaptive Detail

Blake's study aims at reducing the computational complexity of computer animation by computing just what is needed to produce convincing pictures. The thesis is centered around two metrics: a spatial (static) priority–"objects further away from the viewpoint are visually less important to the picture than those closer by"– and a temporal (dynamic) priority–"objects moving quickly with respect to the observer need to be redrawn more often than those at rest". These initial intuitive formulations of the two metrics are further developed and extended with the notion that human vision introduces a trade-off between temporal and spatial resolution. These are the basics for arriving at measures to determine to what extent detail can be left out without being noticed by a viewer.
Changing images are a function of two space variables and of time. A Fourier transformation of this function results in a spectrum, a function of temporal and spatial frequencies. This function is non-zero in a limited domain only. The effect of motion on this spectrum is that it is sheared in the temporal frequency dimension. The amount of shear is proportional to the velocity of the object components in the two dimensional image.
The human vision system is limited in spatial and temporal frequencies by the various transmission systems between our mind and the outside world, resulting in a so-called window of visibility [Watson 1986]. Since the spectrum of moving objects is sheared, some spatial frequencies which would fall inside the window of visibility for static objects may fall outside this window when these objects are in

motion. Removing the spatial frequencies that fall outside the window of visibility is equivalent with removing invisible detail of the object. However, this is exactly the detail that in combination with the frame rate of the display system appears in aliased form, causing the unnatural or jerky motion mentioned above.
Blake remarked that such Fourier analysis is useful for analyzing a problem and obtaining insight into what is happening. However, the computations and algorithms are less likely to use Fourier techniques directly.
Blake argues that optic flow analysis as introduced by Gibson [Gibson 1979] can be used to obtain a measure of the velocity and/or distortion of object components in the image. This measure can be used to determine the amount of detail needed, in other words, which spatial frequencies have to be left out, i.e., to what extent the image should be motion blurred. For planar objects, the recognition of four orders of optic flow effects results in four orders of frame to frame coherence:

- No relative movement
- Translation in the image plane
- Linear transformation in the image plane (shear, rotation, scaling)
- Non-uniform motion, large distortions

5 Computer Animation Versus Animate Response

The major concern of computer animation is to convince the viewer, to make him see what he is meant to see. In general, images are rendered at the highest quality possible. For real-time animation systems such as flight simulators this requires a major exertion which can be maintained by specialized systems and for pre-defined situations only. Off-line computer animation involves three main tasks: modeling, animation (i.e., the design of the choreography) and rendering. These tasks are iterative design processes. As far as these design processes are supported by an interactive editor, the interactivity involves just a minor facet of the animation. This type of animation production is a computationally intensive process which in general takes orders of magnitude more time than the real-time display of the animation itself.
This is quite different from what is needed in an interactive environment of which the main feature is, "The quality of the graphics is not extremely important, as long as the response is good". In our situation the main objective is real-time display of continuous motion, even if we have to deal with complex structures. There is not just one but two temporal limits that complicate the real-time display of continuous motion. These limits are the maximum frame rate of the display system and the capacity of computing resources.

6 Conclusion

What we have learned from computer animation is how to act upon temporal limitations as enforced by the display system and the human vision system. We also learned that in computer animation production the limitation of computing resources is traded against production time.
In order to account for limited computing resources, all we can do for animate response is trade against image quality, i.e., simplify the image generation. Unfortunately, we have to simplify the image generation more than just by leaving out detail which is not noticeable by a viewer, yet we like to present the best possible image. From the example mentioned in the introduction–manipulation of a window boundary instead of manipulation of the window itself–we learned that even to this extent a trade between image quality and system response time can often be quite adequate.
For animate response, we would have to have a metric for the level of detail related to the computing resources available. The implicit dependency as presented in Blake's work should be made explicit.
How the limitation of computing resources effects visualization of continuous motion will be discussed in the next section.

7 Visualization of Motion

As became clear in the previous section, for animate response there are two temporal limits that complicate the visualization of continuous motion. One is the limit on the frame rate as dictated by the display system. The other is the limited capacity of computing resources. The impact of these two limitations is the same. The frame rate needed to display continuous motion is not sufficient, so that visible temporal aliasing may occur. [2]

8 Temporal Anti-aliasing

Whenever the frame rate introduces temporal aliasing, can one of the temporal anti-aliasing methods used for computer animation be of help? If we analyze the methods presented in [Korein 1983; Potmesil 1983; Grant 1985] we see that these methods are basically supersampling techniques for which the computational cost of one anti-aliased frame is much more than the cost of an aliased frame. The more efficient method presented in [Max 1985] adds computation cost to each frame.
In case the computing resources can be considered to be unlimited with respect to what is needed by the application, we are in fact dealing with a real-time animation

[2] In fact no matter what discrete frame rate is used, temporal aliasing will always occur. However, it becomes visible only when the alias frequency falls within the window of visibility [Watson 1986].

system. In this situation, we only have to face the limit as set by the maximum frame rate of the display system and the supersampling methods used for computer animation might be useful.
In an interactive environment, however, it is much more likely that complexity of scenes and the real-time requirements turn out to be a combination that makes us face the limitation on computing resources. In this situation we cannot allow any extra computational effort because it would reduce the frame rate even further. It is remarkable that most common temporal anti-aliasing methods used for computer animation production produce images containing less detail (since the high spatial frequencies are filtered out) at a cost higher than the images containing full detail. From this it is clear that these methods are not useful. If we are limited by computing resources, we should have a method that saves rather than adds computational costs.
Reducing the computational costs can be done by reducing the image quality (i.e., complexity), yet we would like to generate the best possible image. In order to do this, we need an adaptive image generator.

9 Adaptive Image Generation

In computer graphics, the image generation process always has to be tuned to the specific needs of the application. Even when the image generation is a batch mode process performed by the most powerful supercomputer, certain trade offs have to be made. If we were at all to know how to generate a physically perfect image, it would by far exceed the processing power of any state-of-the-art supercomputer. Because of this, a range of rendering models emerged, each with specific features and each of which tries to approximate a physically correct image at a different level. Based on the requirements of a specific application, one of these rendering models can be selected. The fact that worst case situations also have to be considered quite often leads to a choice which in practice is far from optimal.
A more sophisticated approach that avoids this problem is adaptive image generation. By adaptive image generation, the application is able to adjust the image generating process to the specific needs of a particular moment. Ideally this results in the best possible image at any time. Globally speaking, there are two ways in which the image generation process can be adapted: based on the rendering process or based on the object representation. We will elaborate on this in order to find out to what extent this can be used for animate response.
A method based on an adaptable rendering process was suggested by Forrest [Forrest 1985]. Rendering of primitives should be supported at different quality levels. As an example, Forrest recognizes five quality levels to draw a line, starting with an aliased Bresenham line drawing, up to a perfect anti-aliased line of which even the line ends can be specified (e.g., rounded or square). He suggests that such a hierarchy in quality can be exploited in the context of personal workstations. Images should first be rendered at the lowest quality level to get the fastest response and

upgraded to higher quality levels if the user does not take immediate action. Another example of a method based on adapting the rendering process is image rendering by adaptive refinement [Bergman 1986]. Similar to what Forrest suggested, this method improves the quality of a static image as long as there is time to do so. The image successively goes through the following phases: display vertices of polygons, display edges of polygons, display flat shaded polygons, add shadowing, display Gouraud shaded polygons, display Phong shaded polygons and finally anti-alias the image where needed. The performance of the method is enhanced by having each phase make use of results of the previous phase and by trimming the data, i.e., selecting which of the polygons should be handled by the next phase.

The significance of these methods is the combination of interactive response and the generation of the best possible image by exploiting different levels of rendering. Both methods however, become effective only in a static situation, making use of otherwise idle cycles in a personal computer. Since the rendering always starts at the lowest level, it is likely that when displaying continuous motion these methods will produce images of the lowest level only. Also, rendering of a certain level will at least partly undo the results of previous levels. For instance, each phase will have to overwrite pixels resulting from previous phases. At a time-scale of a few seconds, this may not be a point of concern, but at frame rates needed to display continuous motion, the actual writing of pixels becomes a dominant factor. Therefore these methods could be made more efficient if we had a way to determine which level of rendering can be supported from the start.

Methods based on object representation adapt the rendering costs by adapting the complexity of the scene description. Hierarchical data structures [Clark 1976], in which sub-hierarchies contain objects modelled in greater and greater detail, are used to render objects at different levels of detail. Such hierarchical structures allow for a simple incremental approach: as long as there is time, objects can progressively be refined. This can be improved by combining it with the metric described by Blake [Blake 1989]. This metric, which takes into account the viewing distance and speed of the objects, can be used to indicate which objects are best candidates for further refinement. The effectiveness of this type of adaptive image generation is quite dependent on the hierarchical data structure, since properties of the structure determine to what extent increments can make use of previously obtained results.

Adapting the complexity of the scene description seems to be best suited for an incremental image generation method. In the next section we will focus on hierarchical data structures that can be used for this purpose. We note, however, that ultimately a combination of adaptive scene complexity and adaptive rendering should be considered.

10 Hierarchical Data Structures

Since we have to be able to deal with limited computing resources, we have to manage the image generation costs. This managing can be obtained by exploiting hierarchical data structures. Hierarchical data structures allow definition of a so-called graphical working set [Clark 1976] which is that fraction of the structure that at a certain time [Hegron 1987] is potentially of interest. This notion has been practiced in flight simulator applications for a long time. There the most appropriate representation of an object is selected from a hierarchical data structure at display time. Such data structures are carefully optimized for the application.
Numerous hierarchical data structures in which objects are modeled in greater and greater detail have been proposed. To select candidate data structures the following issues should be considered: automatic generation of the hierarchy, motion, relation between level of detail and the cost of rendering, temporal anti-aliasing and texturing.

10.1 Automatic Generation

Not all object representations have an inherent hierarchical structure. In such a situation we need a hierarchy that can automatically be generated. Recursive spatial subdivision seems an obvious way to automatically generate an oct-tree like hierarchical structure. Although the result is a simple uniform representation, rendering of such a hierarchy is not very efficient and the representation is not compact. Furthermore, transformation of objects will require restructuring of the hierarchy. As a result such hierarchies are not particularly suitable for animate response.
Rubin and Whitted state: "Creation of a hierarchical database is a non-trivial operation" [Rubin 1980]. They present a homogeneous representation from which nodes of the hierarchy are procedurally generated. Their representation is a graph-structured hierarchy of only bounding volumes. Each bounding volume contains subspaces, which are the bounding volumes of the next level in the hierarchy. Their representation allows sharing of subspaces. Increment of one level in a hierarchy of bounding volumes implies that a volume, being the difference between the bounding volume and its subspaces, has to be subtracted. This subtraction would reveal previously obscured parts of the scene, which would imply a redo of (a part of) the visibility calculation. It would be much more efficient to have a hierarchy that incrementally adds volumes, leaving the rest as it was. The proposed sharing of subspaces greatly reduces the amount of data storage but introduces combinatorial problems when creating the hierarchy. Since an optimal result requires careful consideration, Rubin and Whitted conclude that it would benefit to off-load the structuring of the hierarchy to the model creation stage. Rubin [Rubin 1982] expands the representation to allow for geometrical transformations between nodes. Such an expansion is extremely welcome in an animate environment because it facilitates the use of the hierarchy in representing moving objects.

10.2 Detail versus Cost

In order to optimally balance image quality and temporal aspects of the animation the level of detail should not only be based on the visual aspects but also on the actual costs of the image generation. For implicit animation in particular it would be desirable to have an absolute quantification of the rendering cost as function of the level of detail. With this we could determine the maximum level of detail for which the animation can be sustained. To obtain an absolute quantification of the costs, however, would require an extensive benchmark of the actual rendering hardware used. Another complication that has to be dealt with is that the complexity of the scene is likely to be view dependent. How to obtain such an absolute hardware dependent quantification and how to embody this knowledge in the hierarchy are questions that remain open.

A requisite of a data structure in case absolute quantification cannot be supplied is that the cost to render objects should be more or less proportional to the level in the hierarchy. Adding one level of detail should not undo all results obtained so far, but instead make optimal use of previous calculations. Ideally, rendering by incremental addition of subsequent levels should at most be as expensive as directly rendering at that particular level of detail.

10.3 Temporal Anti-Aliasing

How well can data structures which contain object descriptions in levels of detail support temporal anti-aliasing? As we saw above, temporal anti-aliasing is equivalent to removing high spatial frequencies above a certain cut-off frequency. This cut-off frequency is inversely proportional to the speed of an object. An object can be described by a sum of band-limited terms of increasing frequencies, a representation which can for instance be obtained by Fourier analysis. The effect of temporal anti-aliasing on such representation would be that the faster this object moves, the fewer terms would remain. This behavior may seem ideal since then the rendering costs would be inversely proportional to the speed of the object. Unfortunately an object description by a sum of band-limited terms is in general not the most efficient. It is not a natural representation of objects, so that Fourier transformation is needed. Also the rendering of such a representation is complicated. In a static or near to static situation, a description of even the simplest object would at least need terms up to a frequency equivalent with the display resolution, otherwise visible detail would be lost. As a result, whether a description by band-limited terms pays off is dependent on the extent to which Fourier transformation simplifies temporal anti-aliasing and complicates rendering. It is likely to pay off only in situations where any other representation would be of similar complexity, such as may be the case for textures.

10.4 Textures

For textures expressed as a sum of band-limited terms, Norton et al [Norton 1982] describe a method of limiting texture detail by "clamping" those terms that exceed a certain frequency. Since the authors only consider spatial anti-aliasing, their clamping frequency is fixed as it is determined by the display resolution only. This method can be extended for temporal anti-aliasing if the clamping frequency is related to the speed of the object. Other spatial anti-aliasing methods for textures exploit hierarchies of various resolution texture maps [Crow 1984; Glassner 1986]. Temporal anti-aliasing of such textures can be done by selecting the appropriate resolution map, again dependent on the speed of the object. However, for animate response, incremental image generation is of more concern than temporal anti-aliasing. It may be clear that in this sense textures expressed as a sum of band-limited terms are favorable, since they are inherently incremental. This is opposed to textures represented by various resolution maps, since the costs of texture mapping is independent of the texture density, so that reducing texture resolution will not reduce the cost of the image generation.

10.5 Summary

We conclude that at this moment there is no unified structure which serves all needs. We noted that for rendering, optimal structuring is a requisite so that automatic generation of hierarchies is not likely to be a real-time process. Procedural hierarchies appear to be very attractive, especially when they feature sharing of sub-hierarchies and transformation between nodes. Since obtaining an absolute measure of the cost of rendering is a problem, a requisite of a data structure is that increment of one level in the hierarchy should result in a simple incremental calculation. We did not find an efficient structure that does support temporal anti-aliasing in a general way. A special situation is representation of textures by band-limited terms. This representation can easily be temporal anti-aliased and also has the advantage that rendering costs relates to detail.

11 Research Topics

From the above, it seems that from a technical point of view, basic knowledge and techniques to implement a human computer interface which supports animate response are already there. However, from a human perceptual point of view, adequate knowledge of how to exploit cognitive skills is still lacking. This knowledge is especially of interest for implicit animation, which also incorporates more abstract types of response. The more abstract type of response leads to interesting issues such as the concepts of illusion, the mechanism by which users are able to associate cues with their own experienc,e and the issue of appropriate abstractions.
For this we would have to develop a human computer interaction environment which forms the basis for further research on implicit animation. At first we ought to

restrict ourselves to an interactive system for dynamic exploration and manipulation (which includes editing) of inherently static models. The implementation would be targeted at future low cost workstations.

The development of the system should emphasize design of data structures. As became clear in the previous section, the internal data representation has to fulfill requirements based on visual and temporal aspects of animation. The data structure should preferably be a procedural hierarchical scene description of which all elements contain information how they should be rendered [Cook 1984].

A straight forward method for temporal anti-aliasing is by simplifying the object representation to such extent that ample time remains for a conventional temporal anti-aliasing method. A data structure has to be found which offers a more general and efficient solution.

It should be noted that the internal data representation should match the hardware characteristics and vice versa. For instance, it makes quite a difference whether or not it is a parallel system and if so whether the system exploits image space or object space parallelism. Vice versa, if incremental calculations and temporal anti-aliasing has to be supported, it is unavoidable that pixels will have to be addressed multiple times and results will have to be accumulated. Multiple access per pixel in a real time environment can only be solved by massive parallelism.

For the external world (i.e., the user) the data representation should be naturally presentable and manipulatable. This is a requirement which potentially conflicts with requirements of the internal data structure. It is not likely that one unified data structure can be found. Object oriented methods, however, provide the necessary data abstraction to be able to hide the internal details of a hierarchy implementation.

The first prototype system should exploit a double buffered frame store. Frames can be swapped at a constant frame rate. The invisible frame should be incrementally improved as long as there is time.

This basic system can be optimized in several ways. A first optimization would be to make use of the metric for computing adaptive detail as proposed by Blake [Blake 1989]. This metric is based on two visual aspects: speed of the object and (weighted) viewing distance. In our case, this metric would not be applied to determine the absolute level of detail at which objects should be rendered, but would be used to give priority to the elements of the graphical working set. This priority will be used to selectively increment the level of detail of those elements that will likely improve the image quality most. A complication that has to be dealt with is that the priority is likely to change during the animation.

Making optimal use of frame coherence is an essential optimization factor in interactive environments. In its simplest form frame coherence can be exploited by partially updating the image. The image has to be redrawn only in the area where the image is known to be changed. A more elaborate way to exploit this time based coherence is by manipulating the frame rate. As we saw in the above, the frame rate needed to display continuous motion should be related to the speed of the

object components in the image. Since a movie projector is operated completely time independent of the camera, it is obvious that the simplest way to synchronize both mechanical devices–by operating at a constant frame rate–is chosen. In our situation, however, we do not have to deal with mechanical devices and there is a tight coupling between the image generation system and the image display system so that synchronization is relatively easy. This gives us the opportunity to vary the frame rate, related to the speed of the object components in the image. Even more so, each object might have its individual frame rate.
As we saw in [Blake 1989], the use of frame coherence does not have to be restricted to frames or parts of frames that do not change in time. Certain types of uniform 3-D motion can be approximated in a mere translation or a linear transformation (shear, rotation and scaling) of the object components in the image plane. Exploiting this more complex form of frame coherence might be worth considering. It should be noted however that in general 3-D motion has effect on shading. This will restrict the extent to which such translation or transformation of images can be used.
An incremental system as described is self adapting. Given the real-time requirements it automatically generates images of the maximum obtainable quality level. A less desirable side effect of this solution is that varying complexity of the scene during the animation may cause disturbing switches of level of detail. Since for implicit animation time restrictions are somewhat liberal, we can trade time for detail. It needs to be investigated to what extent it is acceptable to adjust the temporal behavior of the animation (change speed) to avoid disturbing changes of level of detail.

12 Conclusion

Ideally a graphical user interface should be a real-time animation system. However, we saw that limitations of computing resources enforce an approach by which image quality is adaptively reduced to be able to fulfill real-time requirements. To make this possible in a sufficiently flexible way, the need for hierarchical data structures was recognized. Such an hierarchical data structure should on one hand mask and on the other hand exploit the hardware characteristics. Although the basic concepts of such structures are known, a unified structure which satisfies both internal as well as external requirements could not be found.

13 Acknowledgement

The author gratefully acknowledges the useful comments and suggestions given by E.H. Blake.

14 References

[Bergman1986] Bergman, L., Fuchs, and H., Grant, E., "Image Rendering by Adaptive Refinement", Computer Graphics (20, 4) July 1989 pp 29–37

[Blake1889] Blake, E. H., "Complexity in Natural Scenes: A Viewer Centered Metric for Computing Adaptive Detail", PhD Thesis. Queen Mary College, London 1989

[Clark1976] Clark, J. H., "Hierarchical Geometry Model for Visible Surface Algorithms", Comm. of the ACM (19, 10) October 1976 pp 547–554

[Cook1984] Cook, R. L., "Shade Trees", Computer Graphics (18, 3) July 1984 pp 223–231

[Crow1984] Crow, F. C., "Summed-Area Tables for Texture Mapping", Computer Graphics (18, 3) July 1984 pp 207–212

[Fisher1989] Fisher, S., Panel "Virtual Environments and Interactivity: Windows to the Future" SIGGRAPH '89 Boston, August 1989

[Forrest1985] Forrest, A. R., "Antialiasing in Practice", in NATO ASI Series, editor Earnshaw, R. A. (F17) 1985 pp 113–134

[Gibson 1979] Gibson, J. J. "The Ecological Approach to Visual Perception" Houghton Mifflin co, Boston, 1979

[Glassner1986] Glassner, A. S. "Adaptive precision in Texture Mapping", Computer Graphics (20, 4) July 1986 pp 297–306

[Grant1985] Grant, C. W., "Integrated Analytic Spatial and Temporal Anti-Aliasing for Polyhedra in 4-Space", Computer Graphics (19, 3) July 1985 pp 79–84

[Hegron1987] Hegron, G., "Dynamic Management of 3D Scenes", EUROGRAPHICS '87, Editor Marechal, G. August 1987 pp 529–542

[Korein1983] Korein, J., and Badler, N., "Tlemporal Anti-Aliasing in Computer Generated Animation", Computer Graphics (17, 3) July 1983 pp 377–388

[Max1985] Max, N. L., and Lerner, D. M. "A Two-and-a-Half-D Motion-Blur Algorithm", Computer Graphics (19, 3) July 1985 pp 85–93

[Norton1982] Norton, A., Rockwood, A. P., and Skomolski, P. S., "Clamping: A Method of Anti-Aliasing Textured Surfaces by Bandwidth Limiting in Object Space", Computer Graphics (16, 3) July 1982 pp 1–8

[Potmesil1983] Potmesil, M., "Modeling Motion Blur in Computer-Generated Images", Computer Graphics (17, 3) July 1983 pp 389–399

[Rubin1980] Rubin, S. M., and Whitted, T., "A 3-Dimensional Representation for Fast Rendering of Complex Scenes", Computer Graphics (14, 3) July 1980 pp 110–116

[Rubin1982] Rubin, S. M., "The Representation and Display of Scenes with a Wide Range of Detail", Computer Graphics and Image Processing (19) 1982 pp 291–298

[Shneiderman1989] Shneiderman, B., "Future Directions for Human-Computer Interaction", Summer School on User Interfaces, Tampere, Finland, June 1989

[Watson1986] Watson, A. B., Ahumada, A. J. (jr), and Farrell, J. E., "Window of Visibility: A Psychophysical Theory of Fidelity in Time-Sampled Visual Motion Displays", J. Opt. Soc. Am. A (3, 3) 1986 pp 300–307

Applications and Comparison of Different Mathematical Methods to Compute Form Factors for "Radiosity Images"

F. Bresciani, P. P. Rinaldi, and F. Tapparo

Abstract

A fundamental problem in synthetic images is to determine a correct light interaction between surfaces in an environment.

The radiosity method is an algorithm based on theories used in thermal engineering that includes effects of reflection between ideal diffuse surfaces. The computation of geometric factors (Form Factors), which describes the energy exchange between surfaces, represents the most expensive part of computational time necessary to generate a synthetic image.

This paper aims to analyze different approaches to compute form factors. Subsequently, we make a numerical comparison of the results and provide the evaluation of the relative errors and of the computational time.

At the end we present some simple images generated with the different approaches discussed in this paper.

1. Introduction

The generation of realistic images needs an accurate and correct simulation concerning light propagation in an environment. The first algorithms for illumination modelling were developed in a very simple way. The results describe light models that don't directly use well founded physical behaviour and reflection models arbitrarily associated to the environment.

Among the illumination methods developed, the most frequently used even today is the Direct Illumination method. This method is largely used in graphic software packages particularly in solid and superficial modellers. Although the speed and number of surfaces which can be processed is impressive because of the simplicity of the computations, the pictures generated aren't too much realistic.

Lately two methods have been used to generate synthetic images: Ray-Tracing and Radiosity. Both methods provide limited global illumination effects but generate very realistic images.

The name ray tracing, makes it clear that this method colors every point of the image tracing the path of the light ray from source to the eye of the observer. Conceptually very simple, the model allows the representation in a very realistic way effects of light interraction with objects like shadows, reflections and transparencies. Some of the disadvantages of the method are a lengthy computation time and the facts that the image obtained is dependent on the point of view and that it is not possible to represent realistically surfaces that reflect in a diffuse way.

To generate images that simulate a real scene with realism it is necessary to model the physical behaviour of visible light as it is propagated through an environment. Since the intensity and distribution of light in a scene are governed by energy transport and conservation principles, this must be considered if one wishes to simulate accurately different light sources and materials in the same scene.

The radiosity method is surely the most sophisticated method of realistic representation. Born as a methodology for the determination of the temperatures of surfaces inside an enclosure that exchange heat through irradiation and in a diffuse way [1] it was introduced in computer graphics by Nakamae and Goral to compute in a precise way the interreflection of the light in complex environments.

In fact an illumination model based on energy conservation principles is used to account for all interreflections of light in an environment. The illumination calculations are independent of viewing parameters and can be performed on a wavelength basis rather than the particular RGB channels provided by the phosfors of a display device. The results of the global illumination calculations are used in conjunction with the principles of color science to convert the resulting spectral energy distribution to the RGB primaries of the display device.

In the original formulation of the radiosity method, the environment is discretized in a finite number of small surfaces (patches). For every couple of patches geometrical form factors are calculated which describe the relative energy exchanged. Form factors are thus used to generate a set of simultaneous equations that describe global illumination in the environment.

The advantages reached are several. Among these, the most important are that we obtain an image that is independent of the point of view, and that we get precise information about light energy exchanged between surfaces constituting the environment.

The first part of the paper is a detailed analysis of the hemi-cube method of solving form-factors and it aims to evidence some situations where this approach makes a coarse approximation of the form-factors' exact values, as well as other situations requiring excessive computation time.

In the second part we propose some alternative solutions to calculate form factors and we discuss the results.

2. Radiosity Method

The radiosity method describes an equilibrium energy balance in a closed enviroment. We had started considering all emission and reflection processes as ideal diffuse ones (Lambert). The quantity of light reaching a surface needs a complete description of the geometric relationship between all surfaces. The fundamental relationship of the radiosity method is:

$$\text{Radiosity}_i = \text{Emission}_i + \text{Reflectivity}_i * \int_{\text{env}} \text{Radiosity}_j * \text{FormFactor}_{ij}$$

where:

Quantity	Symbol	Description and unit
Radiosity	B	sum of emitted and reflected energy, the total rate of energy leaving a surface (energy/unit time/unit area)
Emission	E	the rate of energy emitted from a surface (energy/unit time/unit area)
Reflectivity	p	the fraction of incident light which is reflected back into the enclosure (unitless)
Form Factor	F	the fraction of the energy leaving one surface which lands on another surface (unitless)

A simple representation of radiosity can be as follows:

Surface i

$\longleftarrow \sum B_j * F_{ij}$ (total incoming energy)

$\longrightarrow E_i$ (emission)

$\longrightarrow p_i \sum B_j * F_{ij}$ (total reflected energy)

$\Big\} - B_i$ (radiosity)

In practice, the environment is subdivided into discrete surface elements, where for each one we assumed a constant value of radiosity so as to describe the interaction of light energy in the environment. To compute the radiosity values we can use the set of simultaneous equations that can be generated, evaluating radiosity characteristics for every patch (a synonym of discrete surface element) in relation to all other patches constituting the environment.

The i-th equation of the system is given by:

$$[-p_i * F_{i-2} - p_i * F_{i-3} \cdots 1 - p_i * F_{i-i} \cdots - p_i * F_{i-N}] * B_i = E_i \ .$$

It is important to note that it is necessary to form and solve a matrix so structured for each frequency band of interest in order to evaluate the radiosity of every patch. At this moment we used three matrices to represent the radiosity through the primary spectral colours (R, G, B).

Moreover, it is also important to remember that the form factors that we use for evaluating the energy exchange are only a function of geometry and are always independent of any frequency (color) considerations.

3. Hemi-Cube Solution

The fraction of radiation leaving one surface and reaching another is defined as the geometric configuration factor between the two surfaces, because it depends on the

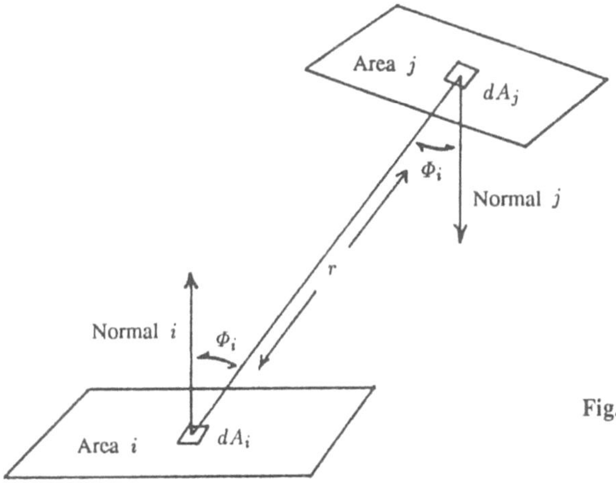

Fig. 1. Geometric form factor

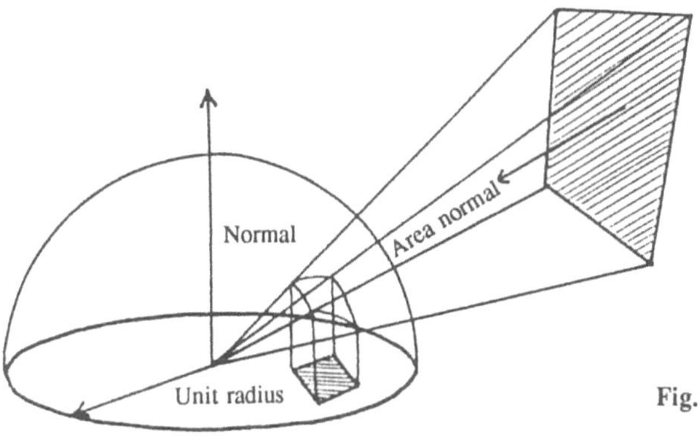

Fig. 2. Nusselt's analogy

geometric orientation of the surfaces with respect to each other and, in this case, it does not depend on viewer position or reflectivity attributes (color) of the surfaces. In fact, form factors which describe the geometrical relationship between any two surfaces are based only on the shape, area, orientation of each surface, the distance between them, and the position of each patch visible to the other. The geometric terms in the form factor derivation are illustrated in Fig. 1.

A geometric analog for the form factor integral was developed by Nusselt: for a finite area, the form factor is equivalent to the fraction of the circle (which is the base of the hemisphere) covered by projecting the area onto the hemisphere and then orthographically down onto the circle (see Fig. 2). If all the patches in the environment are projected onto the hemisphere, the final projection and summations would provide the form factors to all the patches from the patch represented at the center of the hemisphere.

The original idea of Cohen [2] was, instead of projecting onto a sphere, to construct around the center of the receiving patch an imaginary cube with the sides subdivided into small grid cell. With this solution the hemisphere used to compute form factors is replaced with the superior half of the cube and the environment is thus projected on

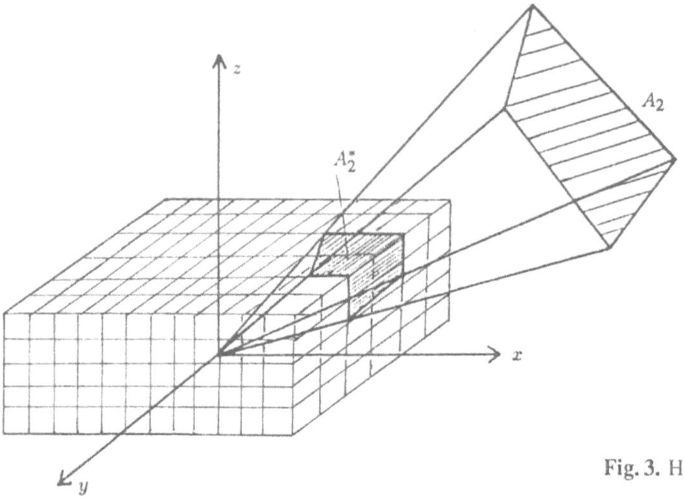

Fig. 3. Hemicube

five surfaces (Fig. 3); in this case a specific delta form factor value for each "pixel" on the hemicube is calculated (found from the equation for the differential area to differential area form factor).

For every patch we have:

$$\Delta F_q = \frac{\cos \Phi_i * \cos \Phi_j}{n * r^2} dA \ , \tag{1}$$

where with dA we indicate the differential area represented by the pixel on the hemicube.

In a concise way we have:

$$F_{ij} = \sum_{q=1}^{R} \Delta F_q \ ,$$

where:

ΔF_q = delta form factor associated with pixel (q) on hemicube

R = number of hemi-cube pixels covered by the projection of patch onto the hemicube.

This procedure was at the base of all radiosity images created in recent years because it intrinsically includes the effects of hidden surfaces.

4. Form-Factor Theory

To make a critical analysis on hemicube solution it is necessary to reexamine the concepts on the base of the form factor theory [3]. One of the chief mathematical complexities in treating radiative transfer between surfaces is accounting for the geometric relations involved in how the surfaces view each other. These effects result mathematically in integrations of the radiative interchange over the finite areas involved in the exchange process. The fraction of energy leaving surface element dA_1 that arrives at element dA_2 is defined as the geometric configuration factor dF_{d1-d2}. We have:

$$dF_{d1-d2} = \frac{dQ_{d1-d2}}{B_1 dA_1} = \frac{B_1(\cos\Phi_1 \cos\Phi_2/\pi r^2) dA_1 dA_2}{B_1 dA_1}$$

$$= \frac{\cos\Phi_1 \cos\Phi_2}{\pi r^2} dA_2 \ , \qquad (2)$$

where $B_1 dA_1$ is the total energy leaving dA_1 within the entire hemispherical solid angle over dA_1. The equation shows that dF_{d1-d2} depends only upon the size of dA_2 and its orientation with respect to dA_1. We can also write the equation in the form:

$$dF_{d1-d2} = \frac{\cos\Phi_1 d\omega_1}{\pi} \ .$$

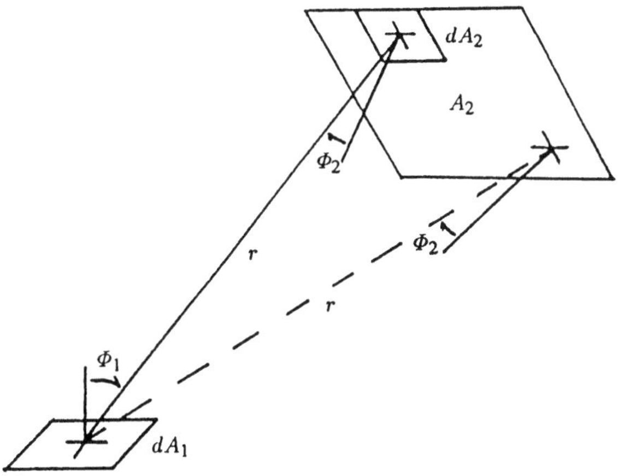

Fig. 4. Radiant interchange between differential element and finite area

Consider now an element dA_1 exchanging energy with a surface of finite area A_2. The relations developed for exchange between differential element must be extended to permit A_2 to be finite. Figure 4 shows (compare the solid and dashed cases) that the angle Φ_2 will be different for different positions on A_2 and that Φ_1 and r will also vary as different differential elements on A_2 are viewed from dA_1. To derive F_{d1-2}, note that the total energy radiated from the surface element dA_1 is $dQ_1 = B_1 dA_1$. The energy reaching dA_2 located on A_2 is:

$$dQ_{d1-d2} = B_1 \frac{\cos\Phi_1 \cos\Phi_2}{\pi r^2} dA_1 dA_2 \ .$$

Then integrating over A_2 to obtain the energy reaching all of A_2 and dividing by the total energy leaving dA_1 results in

$$F_{d1-2} = \frac{\int_{A_2} dQ_{d1-d2}}{dQ_1} = \frac{\int_{A_2} B_1(\cos\Phi_1 \cos\Phi_2 dA_1/\pi r^2)dA_2}{B_1 dA_1}$$

$$= \int_{A_2} \frac{\cos\Phi_1 \cos\Phi_2}{\pi r^2} dA_2 \ , \tag{3}$$

where the integration limits on A_2 extend over only the portion that can be viewed by dA_1. From the equation (2) the quantity inside the integral of (3) is dF_{d1-d2}, so that F_{d1-2} can also be written as:

$$F_{d1-2} = \int_{A_2} dF_{d1-d2} \ .$$

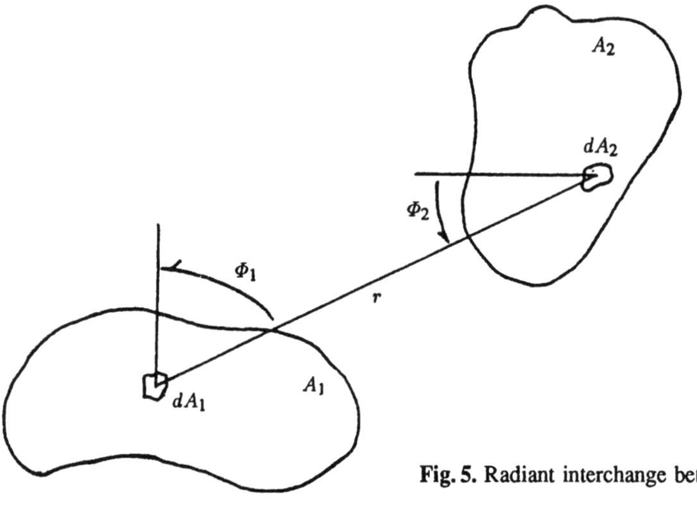

Fig. 5. Radiant interchange between finite areas

Consider the configuration factor for radiation emitted from a surface A_1 shown in Fig. 5 and reaching A_2. By definition, F_{1-2} is the fraction of the energy leaving A_1 that arrives at A_2. The total energy leaving the surface A_1 is $B_1 A_1$. The radiation leaving an element dA_1 that reaches dA_2 was given previously as:

$$dQ_{d1-d2} = B_1 \frac{\cos\Phi_1 \cos\Phi_2}{\pi r^2} dA_1 dA_2 \ .$$

If this is integrated over both A_1 and A_2, then the result will be the energy leaving A_1 that reaches A_2. The configuration factor is then found as:

$$F_{1-2} = \frac{\int_{A_1} \int_{A_2} (B_1 \cos\Phi_1 \cos\Phi_2/\pi r^2) dA_1 dA_2}{B_1 A_1}$$

$$= \frac{1}{A_1} \int_{A_1} \int_{A_2} \frac{\cos\Phi_1 \cos\Phi_2}{\pi r^2} dA_2 dA_1 \ .$$

This can be written in terms of the configuration factors involving differential areas as:

$$F_{1-2} = \frac{1}{A_1}\int_{A_1}\int_{A_2} dF_{d1-d1}dA_1 = \frac{1}{A_1}\int_{A_1} F_{d1-2}dA_1 \ .$$

Analogously we can obtain the form factor formulas where area 2 is receiving and area 1 is emitting:

$$dF_{d2-d1} = \frac{\cos\phi_1 \cos\Phi_2}{\pi r^2} dA_1$$

$$F_{d2-1} = \int_{A_1} \frac{\cos\Phi_1 \cos\Phi_2}{\pi r^2} dA_1$$

$$F_{2-1} = \frac{1}{A_2}\int_{A_2}\int_{A_1} \frac{\cos\Phi_1 \cos\Phi_2}{\pi r^2} dA_1 dA_2 \ .$$

From these formulations of the form factors a reciprocity relationship can be derived for radiosity distributions which are diffuse and uniform over each surface:

$$A_i F_{ij} = A_j F_{ji} \ .$$

A summary of configuration-factor relations is shown in the following table:

Geometry	Configuration factor	Reciprocity
(A) Elemental area to elemental area	$dF_{d1-d2} = \frac{\cos\Phi_1 \cos\Phi_2}{\pi r^2} dA_2$	$dA_1 dF_{d1-d2} = dA_2 dF_{d2-d1}$
(B) Elemental area to finite area	$F_{d1-2} = \int_{A_2} \frac{\cos\Phi_1 \cos\Phi_2}{\pi r^2} dA_2$	$dA_1 F_{d1-2} = A_2 dF_{2-d1}$
(C) Finite area to finite area	$F_{1-2} = \frac{1}{A_1}\int_{A_1}\int_{A_2} \frac{\cos\Phi_1 \cos\Phi_2}{\pi r^2} dA_2 dA_1$	$A_1 F_{1-2} = A_2 F_{2-1}$

Now it is possible to prove that the hemicube algorithm is simply the formulation of form factor between a differential area and a finite one (see case (B) of the table); in fact from the formula:

$$F_{d1-2} = \int_{A_2} \frac{\cos\Phi_1 \cos\Phi_2}{\pi r^2} dA_2$$

using the Nusselt's analogy (referring to Fig. 3):

$$\frac{\cos \Phi_2 dA_2}{\pi r^2} = \frac{\cos^* \Phi_2 dA_2^*}{\pi (r^*)^2}$$

we can obtain:

$$F_{d1-2} = \int_{A_2^*} \frac{\cos \Phi_1 \cos^* \Phi_2}{\pi (r^*)^2} dA_2^*$$

but from formula (1) it is possible to write that:

$$F_{d1-2} = \int_{A_2*} \Delta F_q \simeq \sum_{q=1}^{R} \Delta F_q = F_{ij} \ .$$

From this it is possible to make the first consideration; computing the form factor with the hemicube technique we introduce an error which is proportional to the size of the surface (receiving) of the patch which we consider differential. In the following paragraph we shall make a numerical evaluation of the form factor calculated with the formula (A), (B) and (C) for simple configurations and we shall evaluate the relative errors.

PARALLEL SURFACES

distance	finite_finite	step_step	error (%)	finite_step	error (%)
1,00	2,00E-01	3,18E-01	59,29	2,39E-01	19,83
2,00	6,86E-02	7,96E-02	16,02	7,35E-02	7,13
3,00	3,30E-02	3,54E-02	7,27	3,41E-02	3,44
4,00	1,91E-02	1,99E-02	4,12	1,95E-02	2,00
5,00	1,24E-02	1,27E-02	2,65	1,26E-02	1,30
6,00	8,68E-03	8,84E-03	1,84	8,76E-03	0,91
7,00	6,41E-03	6,50E-03	1,36	6,45E-03	0,67
8,00	4,92E-03	4,97E-03	1,04	4,95E-03	0,52
9,00	3,90E-03	3,93E-03	0,82	3,91E-03	0,41
10,00	3,16E-03	3,18E-03	0,67	3,17E-03	0,33
11,00	2,62E-03	2,63E-03	0,55	2,62E-03	0,27
12,00	2,20E-03	2,21E-03	0,46	2,21E-03	0,23
13,00	1,88E-03	1,88E-03	0,39	1,88E-03	0,20
14,00	1,62E-03	1,62E-03	0,34	1,62E-03	0,17
15,00	1,41E-03	1,42E-03	0,30	1,41E-03	0,15
16,00	1,24E-03	1,24E-03	0,26	1,24E-03	0,13
17,00	1,10E-03	1,10E-03	0,23	1,10E-03	0,12
18,00	9,80E-04	9,82E-04	0,21	9,81E-04	0,10
19,00	8,80E-04	8,82E-04	0,18	8,81E-04	0,09
20,00	7,94E-04	7,96E-04	0,17	7,95E-04	0,08

ORTOGONAL SURFACES

distance	finite_finite	step_step	error (%)	finite_step	error (%)
0,00	2,00E-01	3,18E-01	59,12	1,90E-01	4,95
1,00	3,29E-02	3,54E-01	7,38	3,40E-02	3,09
2,00	1,24E-02	1,27E-02	2,66	1,26E-02	1,25
3,00	6,41E-03	6,50E-03	1,36	6,45E-03	0,66
4,00	3,90E-03	3,93E-03	0,82	3,91E-03	0,40
5,00	2,62E-03	2,63E-03	0,55	2,62E-03	0,27
6,00	1,88E-03	1,88E-03	0,39	1,88E-03	0,20
7,00	1,41E-03	1,42E-03	0,30	1,41E-03	0,15
8,00	1,10E-03	1,10E-03	0,23	1,10E-03	0,11
9,00	8,80E-04	8,82E-04	0,18	8,81E-04	0,09
10,00	7,21E-04	7,22E-04	0,15	7,21E-04	0,08
11,00	6,01E-04	6,02E-04	0,13	6,01E-04	0,06
12,00	5,09E-04	5,09E-04	0,11	5,09E-04	0,05
13,00	4,36E-04	4,37E-04	0,09	4,36E-04	0,05
14,00	3,78E-04	3,78E-04	0,08	3,78E-04	0,04
15,00	3,31E-04	3,31E-04	0,07	3,31E-04	0,03
16,00	2,92E-04	2,92E-04	0,06	2,92E-04	0,03
17,00	2,60E-04	2,60E-04	0,05	2,60E-04	0,03
18,00	2,32E-04	2,33E-04	0,05	2,32E-04	0,02
19,00	2,09E-04	2,09E-04	0,04	2,09E-04	0,02

Fig. 6. Method's comparison

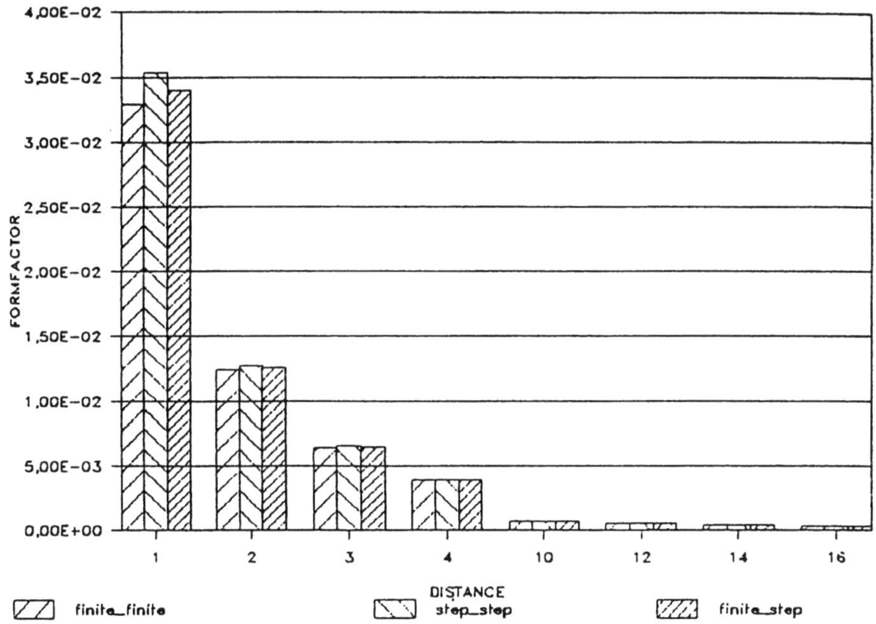

Fig. 7. Orthogonal surfaces

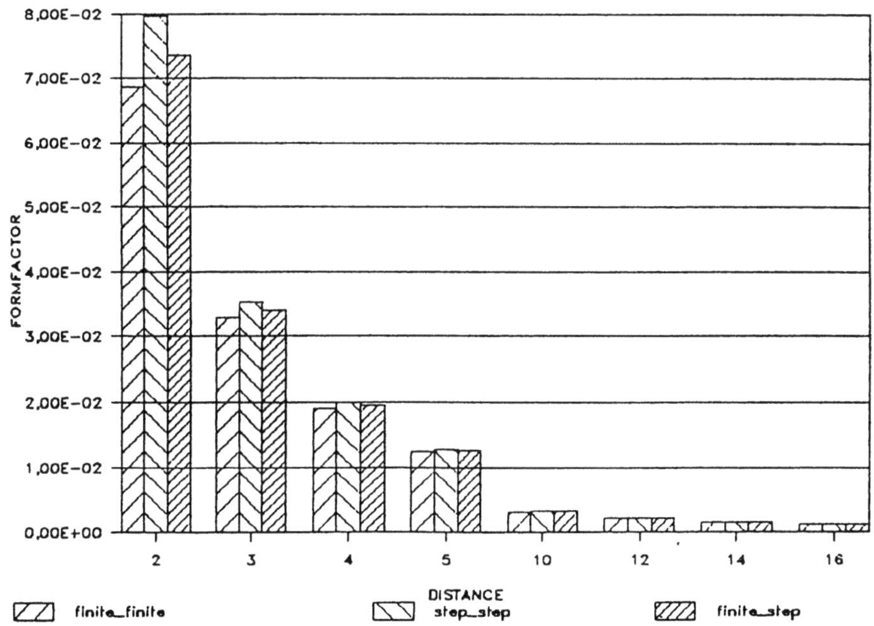

Fig. 8. Parallel surfaces

5. Numerical Evaluation and Relative Errors

We started analyzing two fundamental configurations to evaluate the results of the different methods of form factor calculus. The two fundamental configurations are represented by:

- parallel surfaces landing one in front of each other (area = 1)
- orthogonal surfaces landing one in front of each other (area = 1).

We obtained the relative errors between analytical approach [4] and approximation approaches in that case where the distance was compute among boundary of the patches and the results are given by Fig. 6.

The same results are more interpretable as in Figs. 7 and 8.

6. Applications on Simple Examples

The three methods described in the previous paragraph were implemented in order to visualize a room with a chess floor, thus enabling the researchers to evaluate the effect on the image of the approximation errors made in calculating form-factor. We report in the following table the physical characteristics to compute radiosity values relative to the environment:

Element	Reflectivity			Emissivity		
	R	G	B	R	G	B
Floor	0.95	0.95	0.95	0.0	0.0	0.0
	0.3	0.3	0.3	0.0	0.0	0.0
Walls without door	0.9	0.8	0.55	0.0	0.0	0.0
Wall with door	0.6	0.9	0.9	0.0	0.0	0.0
Door	0.8	0.8	0.8	1.8	1.8	1.8
Cealing	0.85	0.85	0.85	0.0	0.0	0.0
lights	0.8	0.8	0.8	3.0	3.0	3.0

Comparing the methods, relatively to the entire computational time, the results suggest that to the analytical method to compute form-factors corresponds a shorter computational time to solve the energetic system. Consequently, the more approximate the calculation of form factors is, the more computational time will be needed to solve the system (see Figs. 9, 10 and 11).

Moreover, analyzing the three images it is possible to perceive the errors calculated numerically previously.

The error resulting from the form-factor computation considering the differential areas is always positive. The overestimate of the radiosity values, using the same

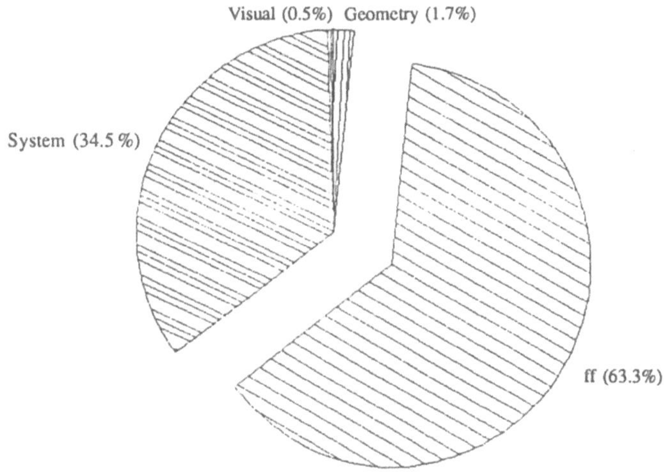

Fig. 9. Computational time. Finite vs. finite

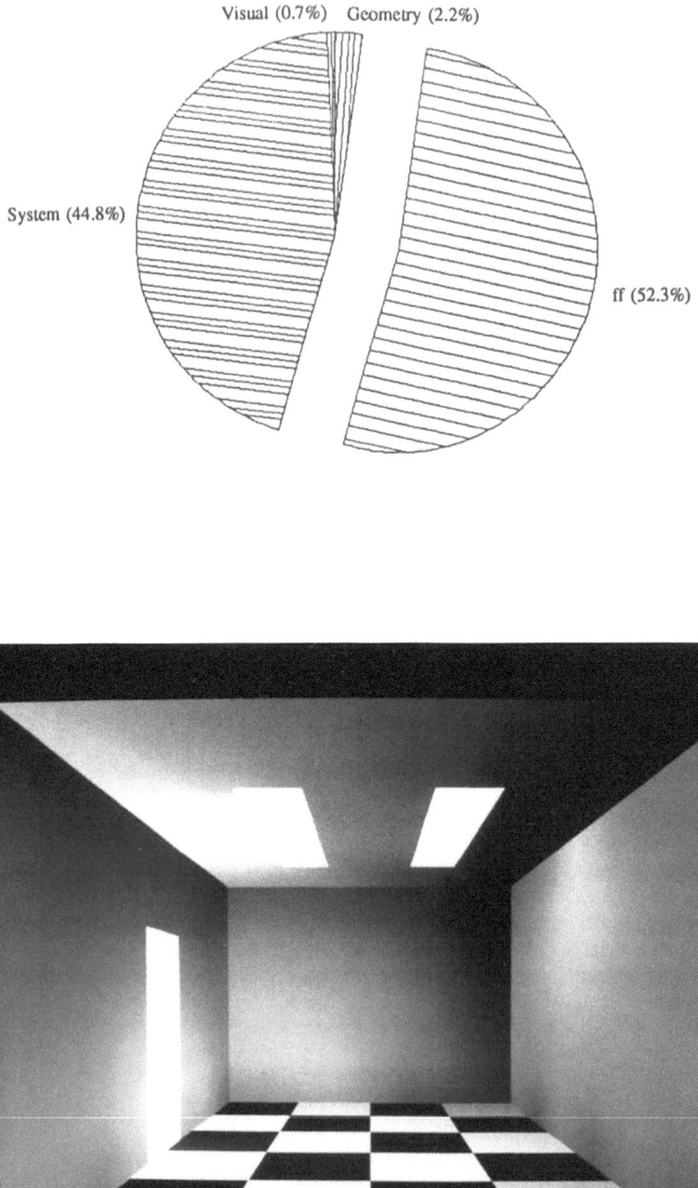

Fig. 10. Computational time. Finite vs. step

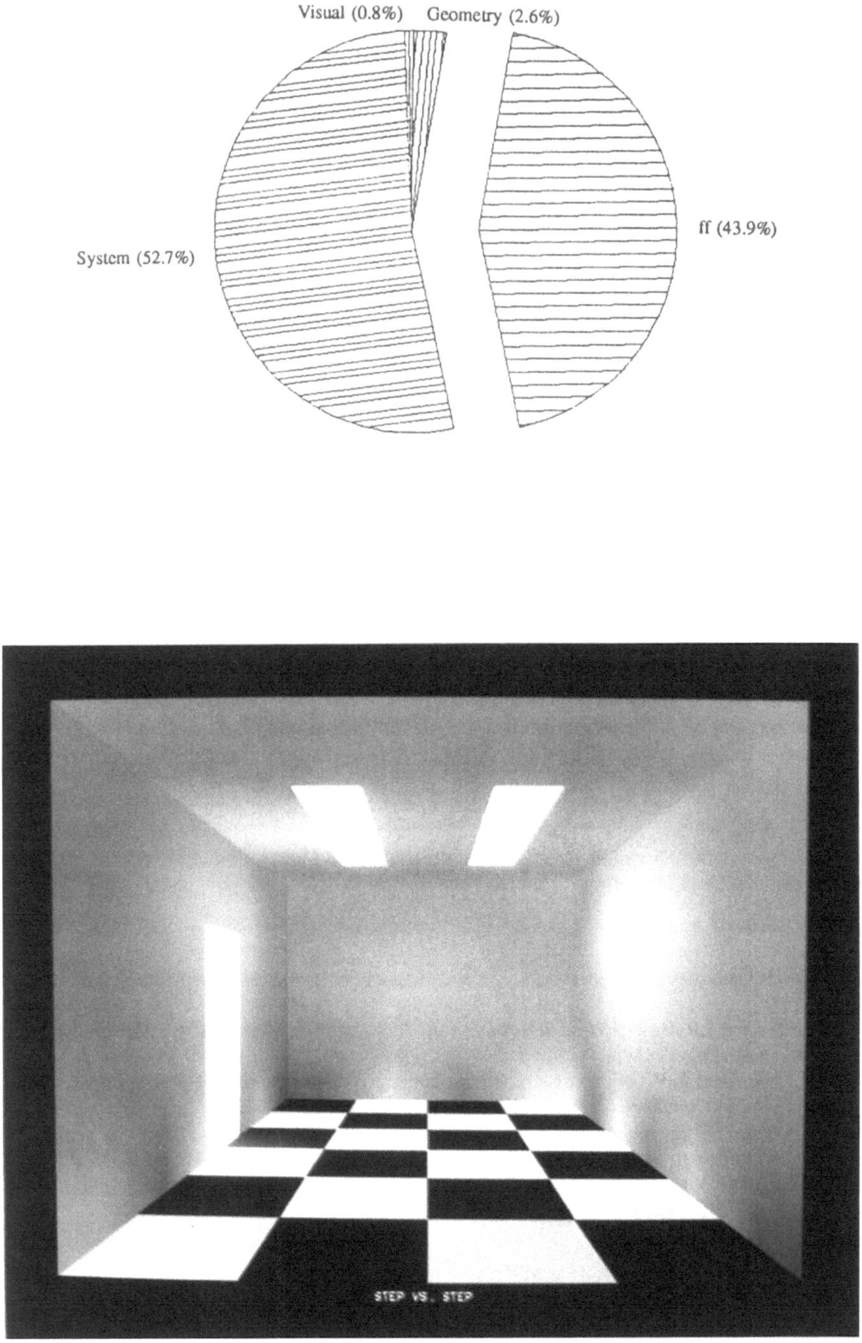

Fig. 11. Computational time. Step vs. step

physical and geometric parameters describing the environment, means a lighter image will be produced.

The difference between the environment analytically compute and the environment compute with the hemicube solution is very interesting; in this case the form factors values are overstimated respect to exact values without the case of adjacent areas ortogonal or parallel surfaces very closed. This fact means to have an image with a light gradient more marked.

Moreover, it is possible to observe that the zones close to the edges are more dark while the far ones are more light.

The last observation is that, without a comparison of this kind, it is practically impossible to understand which of the three environments with identical initial conditions is closer to the real scene.

7. Conclusions

From the analysis of the results we deduced that the technique to compute the form-factors with the hemicube solution is correct when the distance between the patches is at least four times greater than the side of the patch.

For what concerns smaller distances, an error is introduced that is not negligible and this can influence all the generated image.

We believe that there are principally two correct ways to compute form-factors: one solution is to use a mixed method considering areas and relative distance between patches and apply different methods for different situations.

Another approach consists in the definition of a weight function that correct the approximation formula.

Acknowledgements. We would like to thank Prof. Augusto Mazza from Polytechnic of Turin for his comment and support and Hewlett Packard for providing the equipment.

References

1. R. Siegel and J.R. Howell, *Thermal Radiation Heat Transfer*. Hemisphere Publishing Corp., Washington, DC, 1981
2. M.F. Choen and D.P. Greenberg, "The hemicube, a radiosity solution for complex environments", *ACM Computer Graphics* (SIGGRAPH '85) *19*, no. 3, 31–40
3. M.F. Choen, D.. Greenberg, D.S. Immel, and P.J. Broch, "An efficient radiosity approach for realistic image synthesis", *IEEE Computer Graphics and Application 19*, no. 2, 26–33
4. F.M. Sparrow, "A new and simpler formulation for radiative angle factors", *Journal of Heat Transfer*, pp. 81–89, May 1963

Comparative Operations in Solid Modeling

Myeong Won Lee, Toshiaki Satoh* and Tosiyasu L. Kunii

ABSTRACT

Functions for manipulating objects are essential for scientific visualization systems because it is necessary to handle all related data efficiently and to generate objects satisfying given conditions to objects. If operations to compare objects both topologically and geometrically when visualizing the real world, we can model objects more credibly and validly. We introduce comparative operations for this purpose. They are based on a boundary representation incorporating the concept of connected component as a separate geometric entity. Quantitative and qualitative characteristics for all geometric entities which compose objects are analyzed during comparison. The necessary algorithms for performing comparative operations have been developed, and implemented on a commercial solid modeler.

1. INTRODUCTION

Currently, the area of scientific visualization is growing rapidly. A scientific visualization system has to be able to define and manipulate objects in the real world. To do that, it is indispensable to construct a model of the real world. If we model the real world to be composed of sets of objects, the model must incorporate functions for defining and manipulating objects. Current solid modelers have adequate definition capabilities. However, the set of object manipulation functions are still incomplete.

When we consider the functions for manipulating objects for scientific visualization systems, the operations such as display, retrieval, storage, comparison and modification of objects must be included. The system can control the flow of analysis on objects to be displayed if objects generated can be compared to each other.

A scientific visualization system can be integrated with CAD systems and the object modeling facilities of the latter can be made use of. However, restrictions which form part of most CAD systems pose unnecessary limitations.

Most CAD systems generate solids in an interactive manner. They have an advantage of user friendliness but one of their main drawbacks is that tremendous time is required for generation of all solids in a case where the number of solids is very large. When all the solids are to be handled together, an automatic processing mechanism involving control functions is required to generate them at one time. Therefore, a scientific visualization system which incorporates a solid modeler needs to use batch processing to generate many solids at the same time. When batch processing as well as interactive processing for CAD systems is under consideration, the function of manipulation besides definition as in a programming language or a database manipulation language should be also included. The difference is that the manipulation in CAD systems handles solids and the manipulation in a programming language or a database manipulation language handles alphanumeric data. Among the manipulation constructs in a programming language, the **if** statement is important to maintain the integrity of data. It controls the flow of alphanumeric data to reach a solution. If such an **if** statement is available in CAD systems, the flow of solid generation can be controlled to obtain reliable results. This is one motivation for the comparative operations discussed in this paper.

Comparative operations are also required to test the topological and geometrical validity of objects when objects are modeled through several transformations. For example, consider the case where the object that has been transformed is to be retransformed into the original one. The designer may wish to test if the object that has been obtained as the result of retransformation is the same as the original one. Consider another case where many intermediate objects are generated during object design in solid modeling. Comparative operations are required to test whether the intermediate objects are correct or not. Although two objects may be seen to be topologically and geometrically equivalent, they can be changed during a set operation or some other transformation operations.

Consider the case where two objects are composed of several connected components. A connected component denotes a set of loops connected to each other. It indicates also that all the edges in the set of loops are connected. Fig. 1 shows an example object that is composed of four connected components. In this case, each of the corresponding connected components of solids is equivalent in topology and geometry. However the overall geometries of the solids may be different, and then the two solids are not the same after all. This is a difficult problem in an environment using only traditional boundary representation or constructive solid geometry.

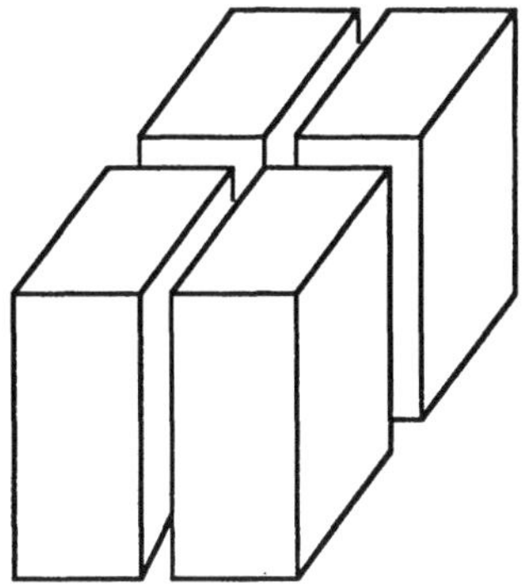

Fig. 1 An object composed of 4 connected components

If quantitative or qualitative characteristics for each object such as length, angle and normal, are maintained, the objects generated can be compared with each other geometrically as well as topologically. In addition, the relationships between geometric entities must also be represented.

This paper presents comparative operations for a boundary representation that can maintain the quantitative and qualitative characteristics for each entity systematically. These characteristics are then used to compare objects both topologically and geometrically. The boundary representation includes a connected component as a geometric entity.The comparison can be done on objects in any location. Therefore the comparative operations are location-independent, or in other words, coordinates-independent and serve as an efficient manipulation facility for real world objects in application such as scientific visualization.

2. A BOUNDARY REPRESENTATION

In general, the geometric entities used for the description of an object in a boundary representation are vertex, edge, loop and solid. We chose loop instead of face so that the concept of outer loop and inner loop can be used. A boundary representation which includes the connected component as an another geometric entity is illustrated.

When an object is composed of several connected components, it is difficult to determine the geometry of the object uniquely using the conventional boundary representations. Our boundary representation can solve the problem by permitting the analysis of the relationships between connected components. It has the structure described in Fig. 2. Using this boundary representation, the topology and the geometry of the objects can be uniquely maintained.

The advantages of our boundary representation are as follows:

- The geometric validity of objects can be easily checked.
- The result of a local transformation on a solid such as translation and rotation of connected component can be easily obtained.

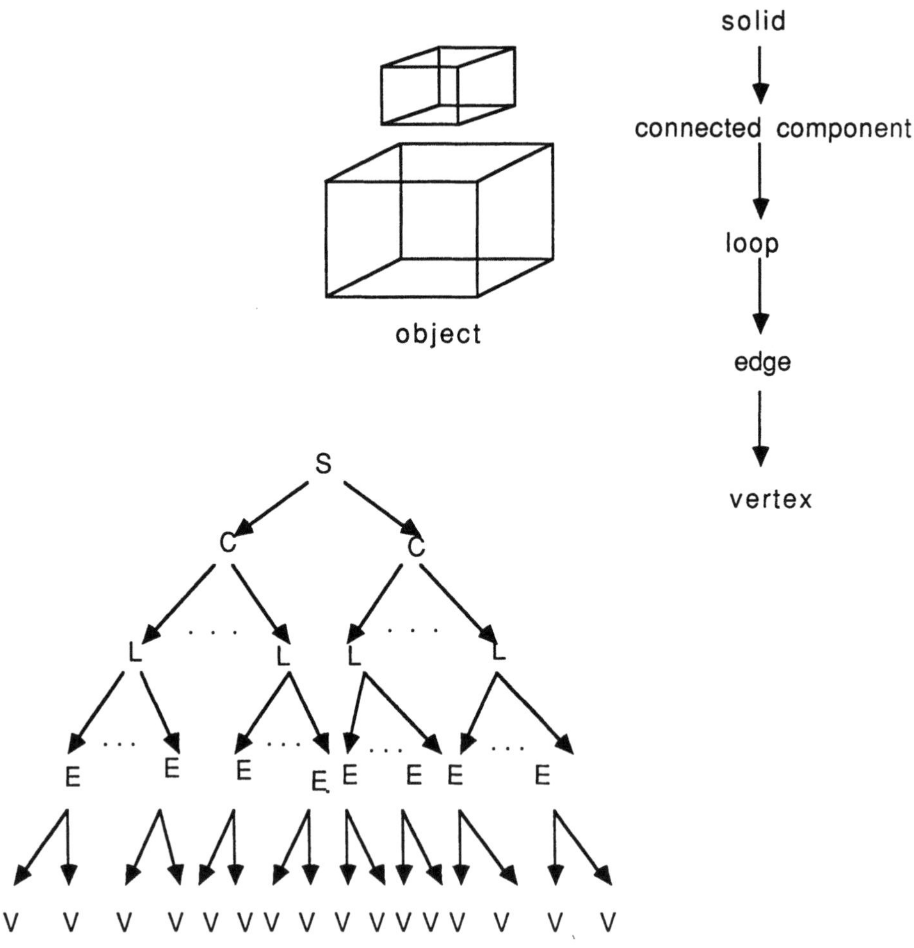

Fig. 2 A boundary representation

Algorithm 2.1 shows the generation of connected components in each object. It is based on FIFO(First In First Out) algorithm by using BFS(Breadth First Search). The algorithm 2.1 starts by inserting initial vertex into queue. All vertices connected to the initial vertex are inserted. Next, the initial vertex is deleted and all vertices connected to next vertex in the queue are inserted. If no more connected vertices exist, all remaining vertices in the queue are deleted. At this point, one connected component has been generated. Then, an unvisited vertex is inserted into the queue to generate another connected component. This process is repeated until all vertices are visited. In the algorithm, **delete** denotes the procedure that deletes the first vertex from a queue. And **insert** denotes the procedure which inserts a vertex into a queue.

Algorithm 2.1:

```
procedure CoCo(CC)
  //CC denotes the array to store the information for connected components //
  //It includes the kinds of connected component and number of loops for each connected
  component//
  begin
  //initialize all elements of array QUEUE to be -1 and insert starting vertex//
  //initialize all elements of array VISITED to be 0//
  //initialize array CCTBL including the kind of connected component for each loop//
  v ← an arbitrary vertex of object
  insert(v)
  KIND ← 1 //KIND denotes the kind of connected component//
  for all vertices w adjacent to v do
     for all loops including the vertex w do
        CCTBL(loop) ← KIND
  loop
     begin
       if QUEUE(0) < 0  then
           begin
             COUNT ← 0
             for all loops do
               if CCTBL(loop) is KIND then  COUNT ← COUNT + 1
             CC(KIND) ←COUNT
             if all vertices have been visited then return
             else
                 begin
                   v ← an unvisited vertex
                   insert(v)
                   KIND ← KIND + 1
                   continue
                 end
           end
       //prepare the next connected component using BFS//
       delete(v)
       if  v is visited then  continue
       VISITED(v) ← 1
       for all vertices w adjacent to v do
           begin
             insert(w)
             for  all loops including the vertex w do
                  CCTBL(loop) ← KIND
           end
     end
```

repeat
end CoCo

3. COMPARATIVE OPERATIONS

Comparative operations can be divided into two categories where one set of operations test for equality and the other for similarity. There are two operators to test equality, equal(=) and not equal(≠). There are two operators to test similarity, greater than (>) and less than (<). Similarity means that the forms of objects to be compared are the same but the sizes are different as shown in Fig. 3.

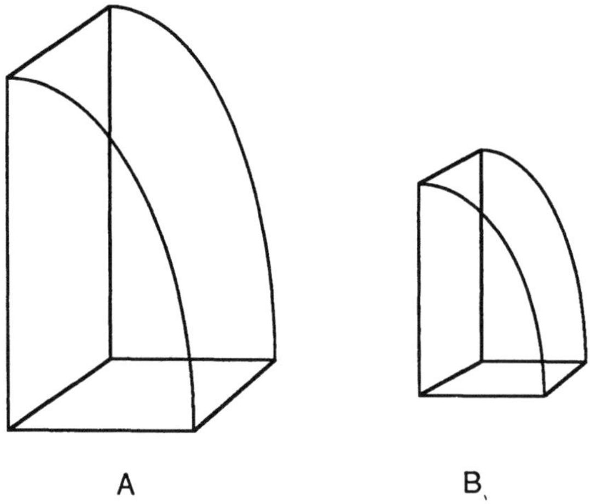

Fig. 3 Similar objects

When two objects to be compared are similar, the greater than (>) operator returns **TRUE** if left object of the operator is bigger than right object. Otherwise, **FALSE** is returned. The less than (<) operator returns **TRUE** if left object of the operator is smaller than right object. Otherwise, **FALSE** is returned. For example, A > B returns **TRUE** and A < B returns **FALSE** in Fig. 3. Operations such as greater than or equal (>=), and less than or equal (<=) can be easily supported by combining equality and similarity operators. These operators can be implemented by analyzing the topological and geometrical properties of objects.

Some previous research exists on the problem of detecting the equality of objects. But they are characterized by some restrictions which are as follows. Tilove[9] discussed that the equality test can be performed by the same method used for null object detection. But it is difficult to compare the objects composed of several connected components using constructive solid geometry. Sugihara's algorithm[6] has the restriction that all faces of an object must be connected. In this case, the problem cannot be solved when an object is composed of multiple connected components. Karasick[8] solved the same object problem by defining a special minimal boundary representation for solids, called B-Rep$_{min}$, a lexicographical ordering for faces and edges. Although two objects are equivalent in geometry as well as topology, Karasick's approach decides that they are not equal if their positions are different in coordinates system. Therefore, the approach is very location-dependent. Turner[5] defined topological validity through the consideration that two topological structures are equivalent if they have the same number of faces, edges and vertices, and if they are connected in the same way. But only topological validity cannot uniquely define an object particularly when an object is composed of several connected components. In addition, all the above methods give rise to problems in the case of the similarity test. For the similarity test, a new algorithm or theory is needed.

Definition 3.1 illustrates the concept of geometrical validity in this paper. It is also required to define an object uniquely besides topological validity which determines the topology of an object valid.

Definition 3.1: geometrical validity

When the objects to be compared are equivalent topologically and geometrically, it is said that they are maintained by geometrical validity.

The comparative operations can recognize the equality of objects irrespective of their location as well as number of connected components they contain. This paper concentrates on the equal operation. Other comparative operations such as 'greater than' and 'less than' can be obtained by similar implementation. Comparative operations test for topological and geometrical validity. The topological validity can be easily tested by analyzing the boundary representation of an object. The geometrical validity can be checked by analyzing the quantitative or qualitative characteristics and the relationships between geometric entities in an object.

Geometric entities such as vertex, edge, loop, connected component and solid have their specific quantitative or qualitative characteristics to represent an object uniquely. These characteristics must be analyzed prior to the implementation of the comparative operations. Before the quantitative and qualitative characteristics are described, the terminology of p-loop and c-loop is illustrated. The terms, p-loop and c-loop[2], are used to represent a relation between loops. When a loop has a hole, the loop is denoted by p-loop and the hole by the c-loop (Fig. 4). In other words, p-loop is the outer boundary of face and c-loop is inner one.

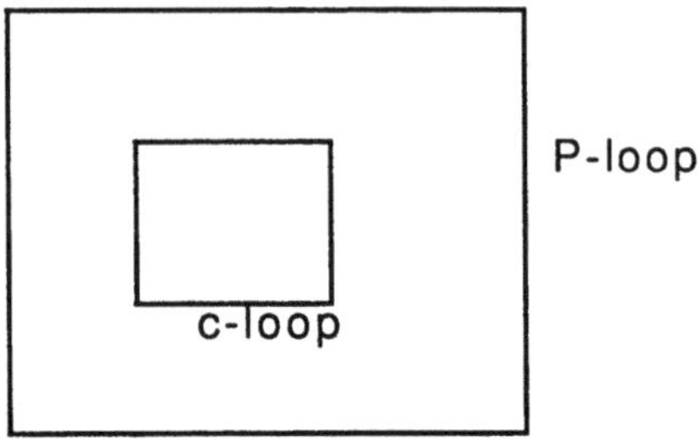

Fig. 4 A p-loop and a c-loop

The quantitative or qualitative characteristics for each entity are as follows:

- the characteristics of an edge:
  length, type (curve or line)
- the characteristics of a loop:
  number of edges, number of curves, number of lines, angle between adjacent edges, normal vector of loop, type of loop (p-loop or c-loop), type of face (plane or curved face), center point
- the characteristics of a connected component:
  number of loops, number of curved faces, number of planes
- quantitative characteristics of a solid:
  number of connected components, relation information between connected components

The relation information has been introduced and analyzed to maintain the validity of objects. The relationships between loops and the relationships between connected components can represent the geometric properties of an object.

4. CONDITIONS FOR GEOMETRIC EQUALITY

The following Definition 4.1, 4.2, 4.3 and 4.4 describe necessary and sufficient conditions to satisfy the geometric equality of objects.

Definition 4.1: conditions for solid equality

Given solids S_1 and S_2, it is concluded that they are equal if the following conditions are satisfied.

condition 1: Two solids S_1 and S_2 are equivalent in topology.

condition 2: All corresponding connected components are equal. This includes that they have the same quantitative and qualitative characteristics.

$$Q(C^1{}_i) \equiv Q(C^2{}_j),\ C^1{}_i \subseteq S_1,\ C^2{}_j \subseteq S_2, \text{ for all i and j},$$

where Q denotes quantitative and qualitative characteristics for a connected component, C denotes a connected component.

condition 3: All corresponding relationships between connected components are the same. These are 3D relationships between connected loops.

$$R(C^1{}_i \rightarrow C^1{}_k) \equiv R(C^2{}_j \rightarrow C^2{}_l),$$
$$C^1{}_i, C^1{}_k \subset S_1 \text{ for all k, } i \neq k,$$
$$C^2{}_j, C^2{}_l \subset S_2 \text{ for all j, } j \neq l,$$

where $R(A \rightarrow B)$ denotes a relationship between A and B, and C denotes a connected component in a solid. $C^1{}_i$ and $C^2{}_j$ must be equal.

Definition 4.2: conditions for connected component equality

Given connected components C_1 and C_2, $C_1 \subset S_1$, $C_2 \subset S_2$, it is concluded that two connected components are equal if the following conditions are satisfied.

condition 1: They are equivalent in topology.

condition 2: All corresponding loops are equal. This includes that all corresponding loops have the same quantitative and qualitative characteristics.

$$Q(L^1{}_i) \equiv Q(L^2{}_j),\ L^1{}_i \subset C_1,\ L^2{}_j \subset C_2, \text{ for all i and j},$$

where Q denotes quantitative and qualitative characteristics, and L denotes a loop.

condition 3: All corresponding relationships between geometric entities composing the connected components are the same. There are 2D relationships between connected edges and 3D relationships between loops.

$$R(E^1{}_i \rightarrow E^1{}_k) \equiv R(E^2{}_j \rightarrow E^2{}_l),$$
$$E^1{}_i, E^1{}_k \subset S_1 \text{ for all k, } i \neq k,$$
$$E^2{}_j, E^2{}_l \subset S_2, \text{ for all l, } j \neq l,$$
$$R(L^1{}_i \rightarrow L^1{}_k) \equiv R(L^2{}_j \rightarrow L^2{}_l),$$
$$L^1{}_i, L^1{}_k \subset S_1 \text{ for all k, } i \neq k,$$
$$L^2{}_j, L^2{}_l \subset S_2, \text{ for all l, } j \neq l,$$

where $R(A \rightarrow B)$ denotes the relationship between A and B, E denotes an edge and L denotes a loop.

Definition 4.3: conditions for loop equality

Given loops L_1 and L_2, $L_1 \subset S_1$, $L_2 \subset S_2$, it is concluded that two loops are equal if the following conditions are satisfied.

condition 1: They are equivalent in topology.

condition 2: All corresponding edges are equal. This represents that corresponding edges composing the loops have the same quantitative and qualitative characteristics.

$$Q(E^1{}_i) \equiv Q(E^2{}_j),\ E^1{}_i \subset L_1,\ E^2{}_j \subset L_2, \text{ for all i and j,}$$

where Q denotes quantitative and qualitative characteristics for an edge in a loop and E denotes an edge in a loop.

condition 3: All corresponding relationships between geometric entities composing the loops are the same. They include the relationships between adjacent edges.

$$R(E^1{}_i \rightarrow E^1{}_k) \equiv R(E^2{}_j \rightarrow E^2{}_l),$$
$$E^1{}_i, E^1{}_k \subset L_1, \text{ for all k,i} \neq \text{k},$$
$$E^2{}_j, E^2{}_l \subset L_2, \text{ for all l, j} \neq \text{l}$$

where $R(A \rightarrow B)$ denotes the relationship between A and B, E denotes an edge and L denotes a loop.

Although the comparative operations can be implemented by using only 3D relationships, the comparison of 2D relationships is also included so that this results in time efficiency in the worst case.

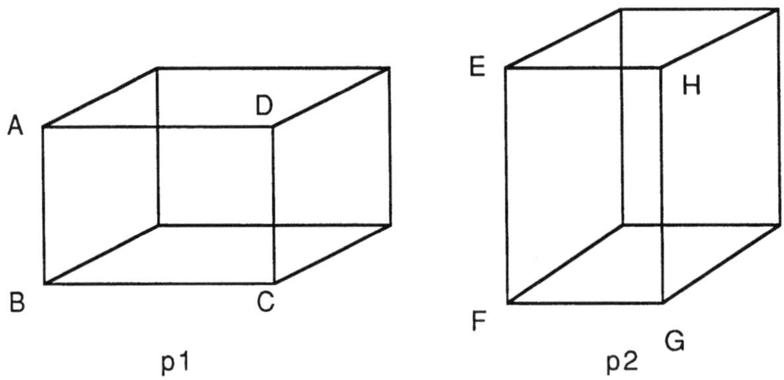

Fig. 5 Loop comparison

When we are to compare two objects as in Fig. 5 that are absolutely equal, corresponding loops are searched, and then four edges and their relationships such as distances and angles between adjacent edges are compared. For example, the corresponding loops (ABCD and EFGH) are equal because four corresponding lengths of edges (AB=EH, BC=EF, CD=FG, DA=GH) and four corresponding angles at vertices are the same (∠ABC=∠HEF, ∠BCD=∠EFG, ∠CDA=∠FGH, ∠DAB=∠GHE). In this case, fewer conditions are sufficient to satisfy the geometric equality of the loops. But we have considered all possible conditions because exceptional cases may occur. If the edges constituting the loop are curved edges, the distances between control points for the curve are compared. In Fig. 6, four lines connecting four control points, A, B, C, D in edge1, and E, F, G, H in edge2, are compared. In addition, corresponding four angles at control points are compared.

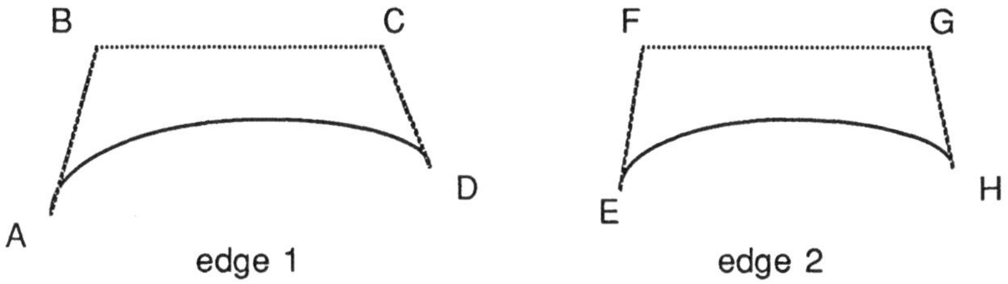

Fig. 6 Curved edge comparison

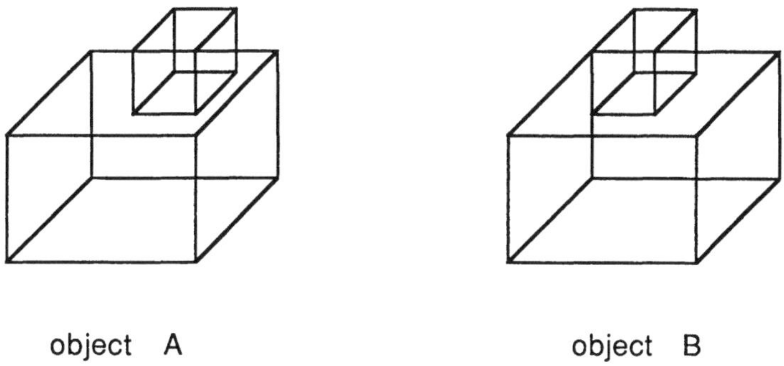

Fig. 7 Objects with the same topology but different geometry

To compare two objects, the following information must be generated for each object.

OBJECT_INFO (object identifier, number of loops, number of edges, number of straight edges, number of curved edges, number of vertices, number of connected components, number of c-loops, number of p-loops)

CONNECTED_COMPONENT_INFO (connected component identifier, number of loops, number of planes, number of curved faces)

LOOP_INFO (loop identifier, type of loop(plane, or curved), number of edges, number of straight edges, number of curved edges, normal vector, angles between adjacent edges)

EDGE_INFO (edge identifier, type of edge(straight or curved), length, vertex identifier)

There are cases that two objects are equivalent in topology, but not equivalent in geometry. Such cases are detected by comparing the relationships between connected edges or between connected loops. In these cases, two important concepts are involved. One is a 2D relationship which defines c-loop uniquely, when the connected components are adjacent as in Fig. 7. There exist unique values for representing the relationship such as the distance and the angle between c-loop and p-loop. In Fig. 8, it cannot be said that the two loops are equivalent in geometry although the corresponding p-loops and c-loops are equivalent. This is because the two c-loops are differently positioned in relation to their p-loops respectively.

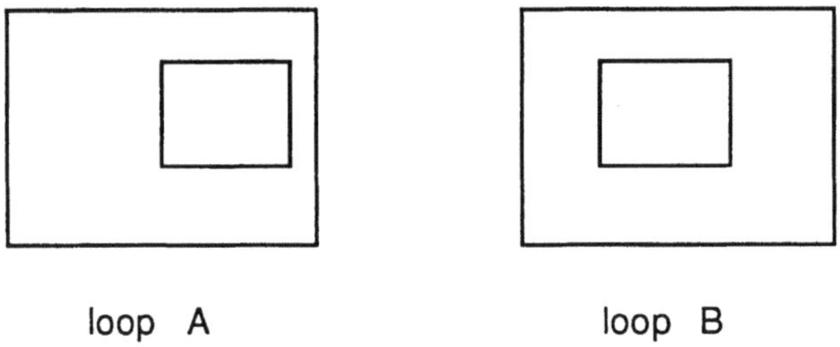

Fig. 8 Differently positioned c-loops

The procedure to calculate the difference in position of c-loops is as follows:

- Select a base edge for each loop.
  These edges should be equivalent in quantitative and qualitative characteristics (length and type).
- Compare the relationships between p-loops and c-loops (distance and angle).

The other concept is a 3D relationship between connected components when connected components are not adjacent such as in Fig. 9. After a set operation or a transformation operation, the resulting object can often be generated in a different form than the one considered. Fig. 9 shows two different objects generated by using an add operation and rotational transformations.

The procedure to compare the relationships between connected components is as follows:

- For each object, select a base loop of a connected component: These two loops must be equal in topology and geometry. Therefore, they must have the same characteristics (number of edges, number of lines, number of curves, angle between adjacent edges, length of edge, type of loop).
- For each loop of all connected components which do not involve the base loop, calculate 3D relationships between the loop and the base loop (the distance and the angle between the loop and the base loop).
- Search for the corresponding connected component and corresponding loop of the other object, and calculate 3D relationships between the loop and the base loop.
- Compare the corresponding relationships.

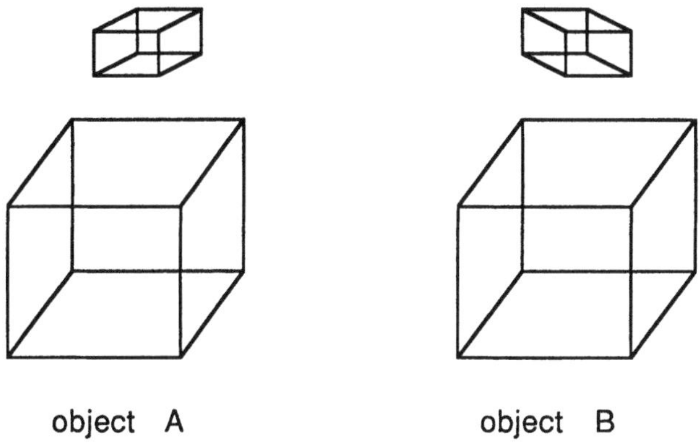

Fig. 9 Objects with different relationships between connected components

5. ALGORITHMS FOR COMPARATIVE OPERATIONS

The comparative operations for objects have been implemented by the following four steps.

step 1: Compare global quantitative or qualitative characteristics of objects.
This involves the total number of connected components, loops, edges and vertices. The number of c-loops and p-loops should also be compared. The total number of straight edges and curved edges should also be compared.

step 2: Compare the lengths of all edges constituting the objects.
For the purpose of efficiency, the edges constituting the object are sorted according to their lengths. Then the corresponding lengths are compared one by one.

step 3: Compare local quantitative characteristics for entities.
The entities involve the connected components, loops and edges. The number of loops are determined for the corresponding connected components for objects being compared. All corresponding loops composing connected components for the objects are compared by their topological properties and characteristics (see Algorithm 5.2 below).

step 4: Compare relationships between entities.
The relationship between two adjacent edges is represented by the angle between the edges. The relationship between p-loop and c-loop is determined by distances and angles between connected edges (see Algorithm 5.3 below). Fig. 10 shows that two objects, A and B, are not equal because the relationships between loops are different. Such a case occurs often when an object is negated in order to accomplish a set operation. It is said to be negated that the normal vectors of all loops are reversed. In this case, the geometrical difference is detected by comparing the relationship between loops which is represented by a distance between center point of one loop and unit normal vector of the other loop (see Algorithm 5.4 below). The center point is obtained by calculating average coordinates for x, y, z respectively. The relationship between connected components is determined by distances and angles (see Algorithm 5.1 below). The distances between connected components are represented by distances between center point of a loop in one connected component and center points of all loops in another connected components. The angles between connected components are represented by angles between unit normal vector of a loop in one connected component and unit normal vectors of all loops in another connected components.

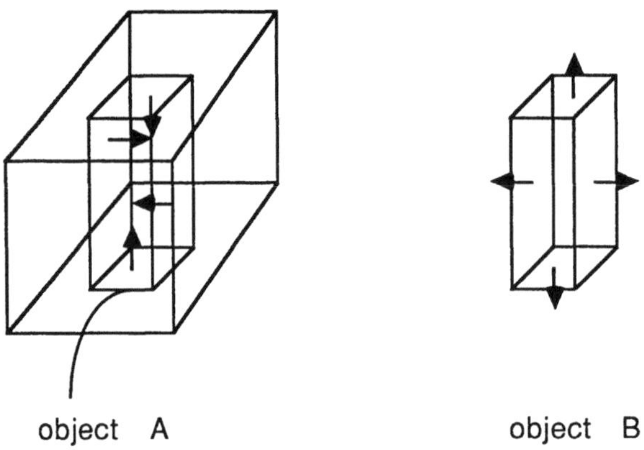

Fig. 10 Different relationships between loops

Algorithm 5.1 represents the procedure RELCOMP that compares the relationships between connected components for two objects.

Algorithm 5.1:

```
procedure RELCOMP(j1, j2)
  //j1, j2 denotes the objects to be compared//
  //prepare the base loop of one connected component of one
  object//
  base1 ← a loop of object j1
  cckind1 ← connected component involving base1
  bnorm1 ← unit normal vector of base1
  childno ← total number of loops involved in all other connected components except
            the connected component involving base1
  COUNT ← 0
  loop
    begin
      //select a loop involved in connected component besides cckind1//
      if all loops have been visited then break
      child1 ←a loop involved in another connected component
      cnorm1 ←unit normal vector of child1
      //calculate the angle between two loops//
      angle1 ←scalar product of bnorm1 and cnorm1
      //calculate the distance between two loops//
      bcen1 ←center coordinates of base1
      ccen1←center coordinates of child1
      dist1 ← distance of bcen1 and ccen1
      //select base loop of j2//
      for all unvisited loops of j2 do
        begin
          base2 ← an unvisited loop of object j2
          if LOOPCOMP(base1,base2) = 'equal'  then break
        end
      cckind2 ←connected component involving base2
      bnorm2 ← unit normal vector of base2
      LOOPCNT ← 0
      loop
        begin
          LOOPCNT ← LOOPCNT + 1
          if LOOPCNT > total number of loops then return ('not equal')
          child2 ← an unvisited loop in another connected component  not involving
                  base loop of object j2
          if LOOPCOMP(child1,child2) = 'equal'  then
            begin
              cnorm2 ← unit normal vector of child2
              //calculate the angle between base2 and child2//
```

```
                angle2 ←scalar product of bnorm2 and cnorm2
                //calculate the distance of base2 and child2//
                bcen2 ← center coordinates of base2
                ccen2 ←center coordinates of child2
                dist2 ← the distance of bcen2 and ccen2
                //compare the angle and the distance//
                if angle1 = angle2 and dist1 = dist2 then
                  begin
                     COUNT ← COUNT + 1
                     if COUNT >= childno then return ('equal')
                  end
             end
          end
        repeat
     end
   repeat
   return ('not equal')
end RELCOMP
```

To compare two loops in topology and geometry, the procedure LOOPCOMP(p, q) has been introduced where p and q denote the loops to be compared. Algorithm 5.2 describes the procedure LOOPCOMP. The function of **adjust** corrects the scalar product according to the direction of edges when the angle between the edges is greater than 180°. Such cases occur when the objects to be compared are convex. Scalar product is used for comparing angles between geometric entities because the correct angles need not be calculated.

Algorithm 5.2:

```
procedure LOOPCOMP(l1, l2)
 //l1 and l2 denote the loops to be compared//
 //j1 denotes the object involving the loop l1//
 //j2 denotes the object involving the loop l2//
 begin
  //initialize//
  line1 ← 0
  line2 ← 0
  curve1← 0
  curve2 ← 0
  //compare number of line and curve//
  for all edges for j1 do
     if e = line edge then line1 ←line1 + 1
     else curve1 ← curve1 + 1
  for all edges for j2 do
     if e = line edge then line2 ← line2 + 1
     else curve2 ← curve2 + 1
```

```
if line1 ≠ line2 then return ('not equal')
if curve1 ≠ curve2 then return ('not equal')
//compare edge length//
sort edge length array LEN1 of loop l1
sort edge length array LEN2 of loop l2
if LEN1 ≠ LEN2 then return ('not equal')
//compare angles at vertices//
for all corresponding vertices for l1 and l2
   begin
     angle1←scalar product of unit vectors of edges at one vertex within l1
     adjust(angle1)
     angle2 ←scalar product of unit vectors of edges at one vertex within l2
     adjust(angle2)
     if angle1≠ angle2 then return ('not equal')
   end
 return ('equal')
end
end LOOPCOMP
```

Algorithm 5.3 represents the procedure to compare c-loops by using the relationship between the c-loop and its p-loop. The c-loops except relationship with p-loops should already have been compared and should be equivalent in topology and geometry.

Algorithm 5.3:

```
procedure CLOOPCOMP(c1,c2)
 //c1 and c2 denote c-loops to be compared//
 begin
  //compare p-loops//
  ploop1 ← p-loop of c1
  ploop2 ← p-loop of c2
  if LOOPCOMP(ploop1,ploop2) = 'not equal'
      then return ('not equal c-loops')
  //compare the relations between c-loop and p-loop//
  for all vertices of c-loop of one object do
    begin
     //initialize//
     DEGEQ ← 0
     pv1 ← one of vertices of ploop1
     pp1 ← adjacent vertex of pv1
     for all vertices of c-loop do
        begin
          cv1 ← one of vertices of c1
          pv2 ← one of vertices of ploop2
          pp2 ← adjacent vertex to pv2
```

```
            cv2 ← one of vertices of c-loop of the other object
            unitv1 ← unit vector of (pp1 – pv1)
            unitv2 ← unit vector of (cv1 – pv1)
            angle1 ← scalar product of unitv1 and unitv2
            adjust(angle1)
            unitv1 ← unit vector of (pp2 –pv2)
            unitv2 ← unit vector of (cv2 –pv2)
            angle2 ← scalar product of unitv1 and unitv2
            adjust(angle2)
            if angle1 = angle2 then
               begin
                  DEGEQ ← DEGEQ + 1
                  if DEGEQ >= number of vertices of c-loop then
                        return ('equal  c-loop')
                  else
                     begin
                        LOOPSHIFT(c1)
                        LOOPSHIFT(c2)
                         continue
                     end
               end
             else
               begin
                  LOOPSHIFT(c2)
                  continue
               end
            if DEGEQ >= number of vertices of c-loop then return ('equal c-loop')
        end
        LOOPSHIFT(ploop2)
     end
   return ('not equal c-loop')
  end
end CLOOPCOMP
```

When the relations between a p-loop and a c-loop are to be obtained, the first thing to do is to search the corresponding vertices of c-loops. In Fig. 11, suppose that the p-loops are equivalent and the c-loops are equivalent. If the relations between the p-loop and c-loop are to be calculated, the base vertex of each p-loop should be selected. If the vertex v1 is selected as the base vertex of p-loop A, the corresponding vertex of p-loop B must be searched. If the lengths of all edges of p-loops are the same, the time required to search the corresponding vertex will be proportional to O(N), where N means the number of vertices in a p-loop. In Algorithm 5.3, LOOPSHIFT is the procedure which selects the adjacent vertex as the next one to be processed if it is found that the current vertex is not the one that corresponds to the vertex in the other object.

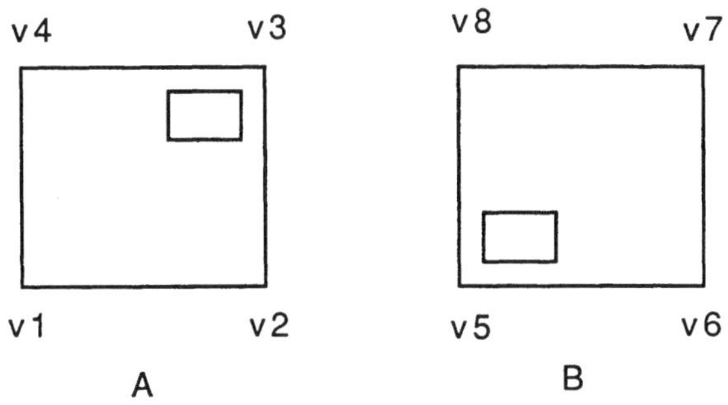

Fig. 11 Searching a base vertex

When set operations are used in solid modeling, the normal vectors of loops may be changed. Consider the case that the difference operation has been executed with the two objects in Fig. 12. The object inside object C is equivalent to object B except the normal vector of each loop. To recognize the difference in their geometries, the 3D relationships between loops should be compared.

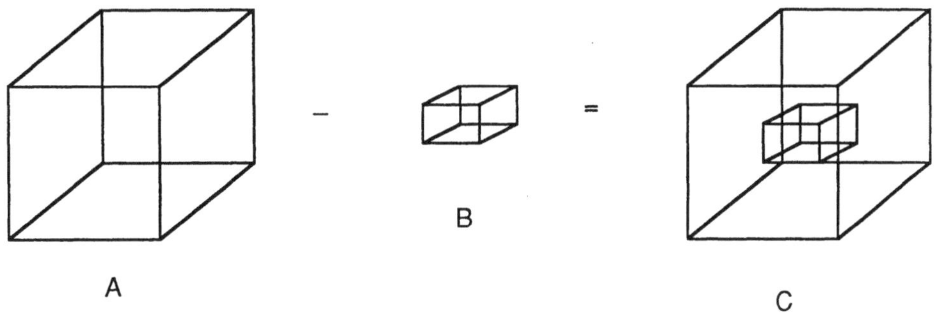

Fig. 12 A set operation

Algorithm 5.4 describes the procedure NEGACOMP for such a comparison. The normal vector of each loop is the most important characteristic. The first thing to do is to select a base loop, which is used for the calculation of the distance between another loop and the base loop. When the distances between the loops are to be compared, the unit normal vector of each loop must be calculated. Then the distance between the normal vector and the center point of base loop is calculated. Hence, the relationships between the base loop and other loops can be obtained. When the objects to be compared are composed of several connected components, the base loop is selected for each connected component and the relationships between the base loop and the other loops of the connected component are compared. If all connected components of the objects have the same relationships besides the same topology, the objects are said to be equal. An important but time-consuming process involves the selection of the corresponding connected components in the objects to be compared.

Algorithm 5.4:

```
procedure NEGACOMP(j1, j2)
 //j1 and j2 denote the objects to be compared//
  begin
   //initialize//
   for all loops of j1 do FINISHED(loop) ← 0
   //select base loop//
   for all connected components do
      begin
       //select a connected component//
       C ← a connected component
       for all loops of C do
          begin
            base1 ← an arbitrary loop of j1
             if FINISHED(base1) = 1 then continue
            cen1 ← center coordinates of base loop
             for all loops of j2 do
                begin
                  base2 ← a loop of j2
                   if LOOPCOMP(base1,base2) ≠ 'equal' then continue
                  //check the loop to be finished or not//
                   if FINISHED(base2) = 1 then continue
                  FINISHED(base2) ← 1
                   break
                end
            cen2 ← center coordinates of base2
            FINISHED(base1) ← 1
             break
          end
        //compare the relationships between base loop and the other loops//
        COUNT ← 0
        for all loops of C do
           begin
             loop1 ← a loop of j1 except base1
             norm1 ← unit normal vector of loop1
             if the connected component including loop1 ≠ C then continue
             for all loops of j2 do
                begin
                  loop2 ← a loop of j2 except base2
```

```
                if LOOPCOMP(loop1,loop2) ≠ equal' then continue
                if FINISHED(loop2) = 1  then continue
                FINISHED(loop2) ←1
                break
              end
            norm2 ← unit normal vector of loop2
            dist1 ← distance between cen1 and norm1
            dist2 ← distance between cen2 and norm2
            if dist1 = dist2  then COUNT ← COUNT + 1
            if COUNT >= (number of loops −1) then
                 return ('equal')
          end
        end
      return ('not equal')
    end
end  NEGACOMP
```

The overall algorithm which compares two objects is described as in the following Algorithm 5.5.

Algorithm 5.5:

```
procedure OBJCOMP(j1,j2)
  //j1 and j2 denote objects to be compared//
  //compare global characteristics//
  loopno1 ← total number of loops of j1
  loopno2 ← total number of loops of j2
  if loopno1 ≠ loopno2  then return('not equal')
  edgeno1 ← total number of edges of j1
  edgeno2 ← total number of edges of j2
  if edgeno1 ≠ edgeno2 then return('not equal')
  vertexno1 ← total number of vertices of j1
  vertexno2 ← total number of vertices of j2
  if vertexno1 ≠ vertexno2 then return('not equal')
  c-loopno1 ←  total number of c-loops of j1
  c-loopno2  ← total number of c-loops of j2
  if c-loopno1 ≠ c-loopno2 then return('not equal')
  lineno1 ← total number of straight edges of j1
  lineno2 ← total number of straight edges of j2
  if lineno1 ≠ lineno2 then return('not equal')
  curveno1 ← total number of curved edges of j1
  curveno2 ← total number of curved edges of j2
  if curveno1 ≠ curveno2 then return('not equal')
  //compare sorted edge lengths//
```

```
LEN1 ← an array of sorted edge-lengths of j1
LEN2 ← an array of sorted edge-lengths of j2
if LEN1 ≠ LEN2 then return('not equal')
for all loops of j1 do VISITED(loop) ← 0 //initialize//
COUNT ← 0
for all loops of j1 and j2 do
   begin
      L1 ← a loop of j1
      if VISITED(L1) = 1 then continue
      L2 ← a loop of j2
      VISITED(L1) ← 1
      if LOOPCOMP(L1, L2) = 'equal' then COUNT ← COUNT + 1
   end
 if COUNT ≠ loopno1 then return('not equal')
 //compare 2D relationships between c-loop and p-loop//
 COUNT = 0
 for all c-loops of j1 and j2 do
    begin
       C1 ← a c-loop of j1
       C2 ← a c-loop of j2
       if CLOOPCOMP(C1,C2) = 'equal' then COUNT ← COUNT + 1
    end
 if COUNT ≠ c-loopno1 then return('not equal')
 //compare 3D relationships between loops//
 if NEGACOMP(j1,j2) ≠ 'equal' then return('not equal')
 //compare 3D relationships between connected components//
 if RELCOMP(j1,j2) ≠ 'equal' then return('not equal')
 return('equal')
end OBJCOMP
```

The comparative operations for geometric equality check have been tested on the objects with varying number of connected components(Fig. 13). Table 5.1 shows the execution time when the objects to be compared are composed of 2, 4, 8, 16, 32 connected components.

6. CONCLUSIONS

The manipulation of objects is indispensable for scientific visualization. Comparative operations have been introduced for this purpose. The operation to detect equal or same objects has been discussed and its implementation has been described in this paper. Other comparative operations such as 'greater than' or 'less than' can also be implemented in the same manner.

The comparative operations are based on the boundary representation which includes the connected component as a geometric entity. Connected components are obtained by FIFO (First In First Out) algorithm with Breadth First Search. The concept of geometric validity has been introduced to compare objects that are composed of multiple connected components. The operation has been implemented using the library facility of a solid modeler DESIGNBASE (ver. 2.0) [4] of Ricoh Co., Ltd. which runs of a VAXstation 2000 with the UNIX operating system.

Case	No. of loops	No. of c. component	Execution time(sec)
A	12	2	1.00
B	20	4	5.90
C	48	8	42.64
D	96	16	156.08
E	192	32	628.74

Table 5.1 Execution time for the comparative operation

ACKNOWLEDGEMENTS

We would like to thank Mr. Yasuto Shirai, Mr. Martin J. Duerst, Ms. Deepa Krishnan and Mr. N. Inamoto for their comments. Our thanks are also due to Dr. Hideko S. Kunii, the director of the SRC for her support to this research.

REFERENCES

[1]Aristides A. G. Requicha, "Representations for Rigid Solids: Theory, Methods, and Systems", Computing Surveys, Vol. 12, No. 4, December 1980, pp437-464

[2]DESIGNBASE Library Manual, Ricoh Company, LTD., 1988

[3]Ellis Horowitz and Sartaj Sahni, Fundamentals of Computer Algorithm, Computer Science Press, Inc., 1978

[4]Hiroaki Chiyokura, Solid Modeling with DESIGNBASE, Theory and Implementation, Addison-Wesley Publishing Company, 1988

[5]Joshua U. Turner, "Accurate Solid Modeling Using Polyhedral Approximations", IEEE Computer Graphics & Applications, May 1988, pp14-28

[6]Kokichi Sugihara, "An *n logn* Algorithm for Determining the Congruity of Polyhedra", Journal of Computer and System Sciences **29**, 1984, pp36-47

[7]Martti Mantyla, An Introduction to Solid Modeling, Computer Science Press, Inc., 1988

[8]Michael Karasick, "The Same-Object Problem for Polyhedral Solids", Computer Vision, Graphics, and Image Processing **46**, 1989, pp22-36

[9]Robert B. Tilove, "A Null-Object Detection Algorithm for Constructive Solid Geometry", Communications of the ACM, Vol. 27, No. 7, July 1984, pp684-694

[10]Silvia Ansaldi, Leila De Floriani and Bianca Falcidieno, "Geometric Modeling of Solid Objects by Using a Face Adjacency Graph Representation", Computer Graphics (SIGGRAPH'85), Vol. 19, No. 3, 1985, pp131-139

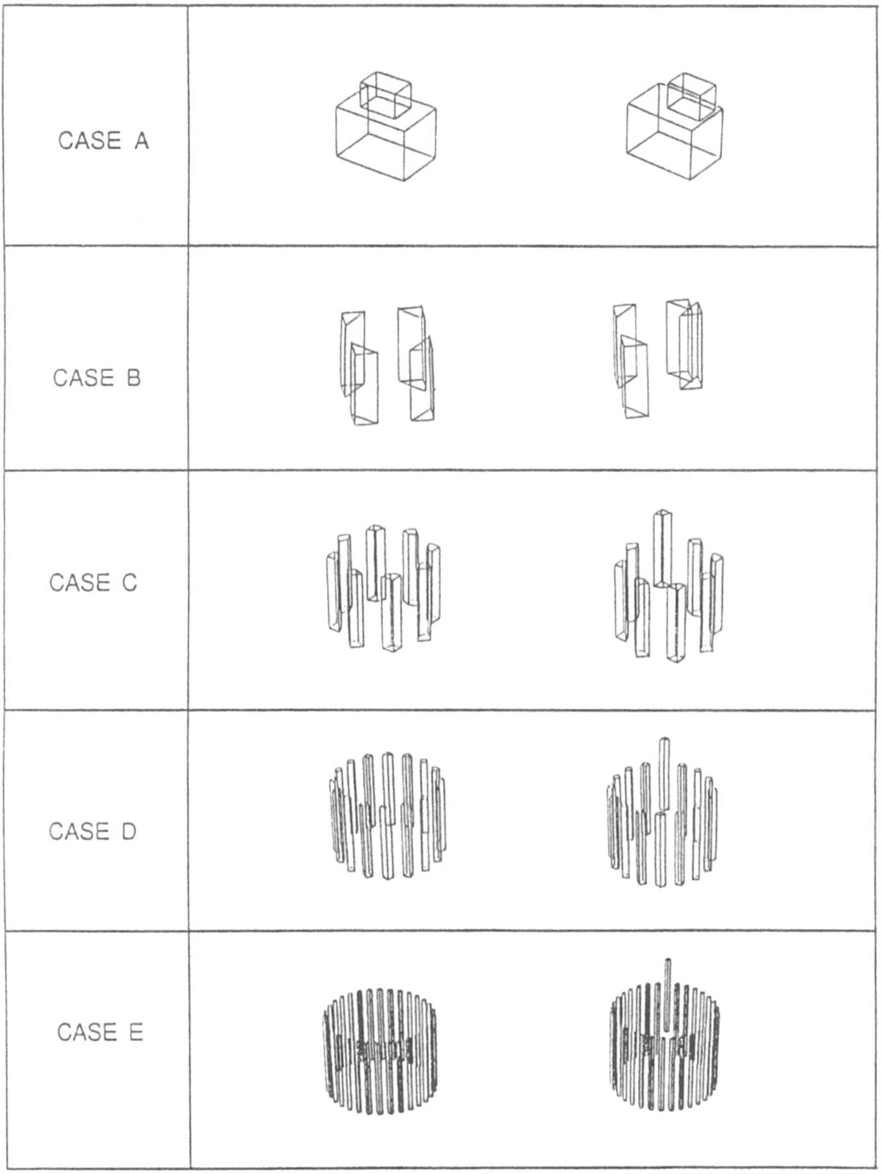

Fig. 13 Test objects with varying number of connected components

3. Highperformance and Multimedia Workstations

3DGRP – A High Performance Graphics System

H. Selzer, T. Haaker, and H. Joseph

Abstract

Modern high performance workstations are based on hardware support for geometry and rendering processing in conjunction with a flexible window management system. These areas are addressed by existing and forthcoming graphic systems and standards like GKS-3D, PHIGS, PHIGS+ and the X- Window System. The 3DGRP is designed to fulfil the different requirements of hardware support in these areas. The architecture presented is a homogeneous solution based on a multi-processor approach with automatic load balancing. The processors are connected tightly to a distributed frame buffer which provides high memory bandwidth. An efficient support of BitBlTs and window addressing mechanisms is given.

1. Introduction and Background

The functional requirements of a graphics system are described best by the classical image generation pipeline (Fig. 1). At the beginning of the rendering process there is the model of a scene. Most often the objects of the scene are represented by a

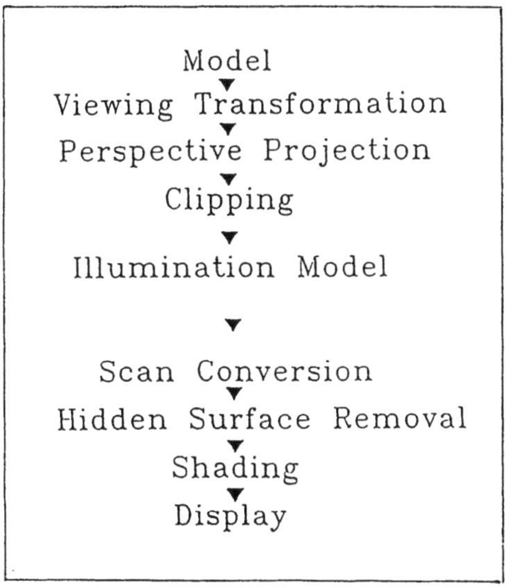

Fig. 1. Image generation pipeline

polygonal approximation. I.e.the graphical objects, which have to be processed by the image generation pipeline, are polygons, defined by vertices, sometimes with color values or normals attached. The viewing transformation and perspective projection are used to choose a specific view of the scene. The transformed graphical objects have to be clipped against the viewing frustum and the workstation window boundaries. In order to generate an image with realistic appearance an illumination model has to be evaluated; i.e. the interaction between different lightsources and the characteristics of the surfaces, as colors, roughness etc., has to be calculated.

The stages just described deal with the geometrical description of the objects, so they are summarized as geometry section of the image generation pipeline.

The following stage which is called scan conversion, maps the geometrical description of an object into the amount of pixels which are covered by that object. The next stage has to solve the hidden surface removal problem, which is realized in modern workstations by the Z-buffer algorithm implemented in firmware. Smooth shading of polygonal objects is accomplished by the algorithms of Gouraud or Phong [ROG85], which can also sometimes be found in hardware or firmware.

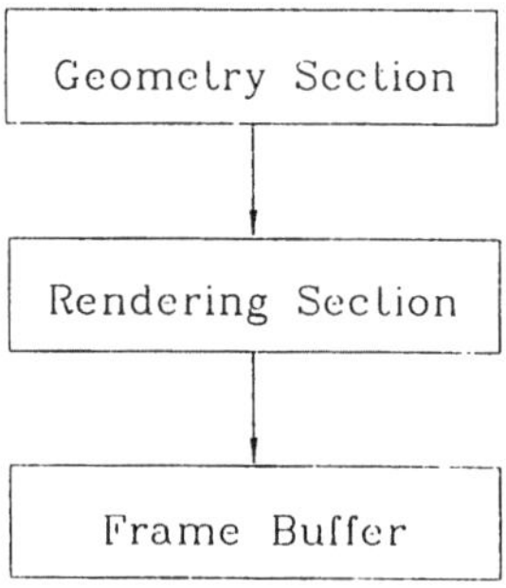

Fig. 2. Image generation scheme

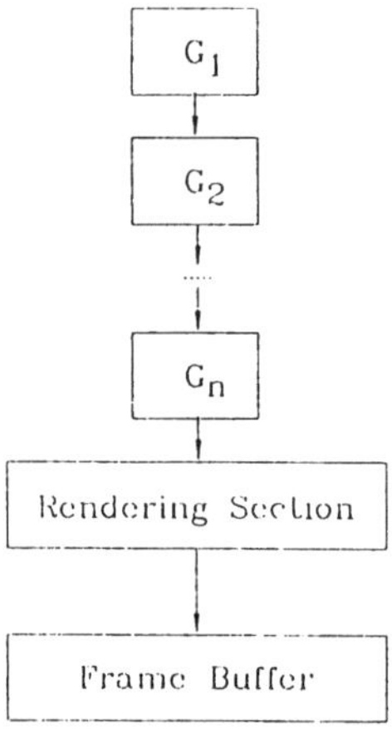

Fig. 3. Geometry pipeline

The last stages of the image generation pipeline are summarized as rendering section. Therefore the image generation pipeline can be reduced to the scheme of Fig. 2. This simple scheme is the basis for most architectures for graphics systems in workstations [TUN87, AKELY88].

In the first time when hardware was developed to speed up graphical algorithms designers concentrated on the geometry section [CLARK82]. As they found a pipeline structure in the algorithms they mapped this structure identically to the hardware, resulting in a geometry pipeline (Fig. 3).

But there are inherent disadvantages of a pipeline structure. First of all it is almost impossible to balance the workload of the geometry pipeline. The transformation of a polygon vertex needs a fixed amount of time because a multiplication of a vector with a matrix has to be performed. But the duration of the clipping process is data dependent. If a polygon edge is totally inside or totally outside the clipping boundaries, the clipping process can be performed very quickly. Otherwise a more lengthy iterative algorithm has to run, to determine where the edge enters or leave the clipping window. That means, the performance of the pipeline can not be fully exploited, because everytime some processors are waiting for some others.

The second disadvantage of pipeline structure is the lack of flexibility. Changing functionality, e.g. clipping at an other number of clipping planes, requires changes in the hardware. On the other hand it is impossible to add processors to the pipeline to increase performance without dramatical changes of the software.

To avoid the problems just described Torborg [TOR87] presents a parallel processing architecture for the geometry section (Fig. 4). The graphical objects are distributed to the processors Gi. Each processor performs all the geometry calculation for an object completely. For all processors Gi are running the same programs, there is no change of software necessary when additional processors are added up. Changing of functionality requires only changing of software but not of hardware.

But even in the global architecture of Fig. 4 there is a two-stage pipeline consisting of geometry and rendering section with all the problems of work load balance. (E.g. assume the processing of a sphere. Then the whole pipeline may be balanced for a certain number of polygons approximating the sphere. Increasing extremely the number of polygons, which cover the same pixels as before, leads to an overloading of the geometry processors while the rendering processors are idle most of the time.)

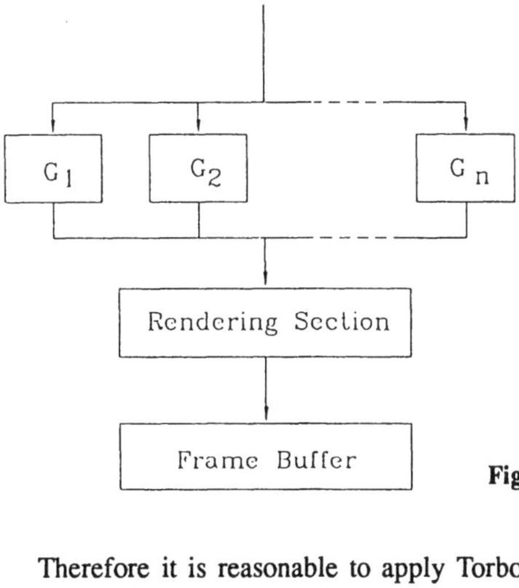

Fig. 4. Parallel geometry processing architecture

Therefore it is reasonable to apply Torborg's approach consequently to geometry and rendering processors. This leads to an architecture consisting of an arbitrary number of non-specialized processors working in parallel, which is realized in 3DGRP.

2.1 Global Architecture of 3DGRP

Realizing the disadvantages of existing architectures the goals for the 3DGRP are as follows:

- *Highly parallel:* To achieve the required system performance.
- *Homogeneous:* Only one type of processor shall be installed, in order to minimize the efforts for hardware and software development.
- *Algorithm-independent:* Changing of algorithms must not require changes in the architecture.
- *Application-independent:* Every application shall be performed with full system performance.
- *Automatic load balancing:* No processor will enter a period of being idle thus achieving the highest possible effectiveness.
- *Easy to extend:* If a performance upgrade is needed the hardware can be extended with minimal effort.

Figure 5 presents the global structure of the 3DGRP. There are a number of identical processors working in parallel and accessing the frame buffer.

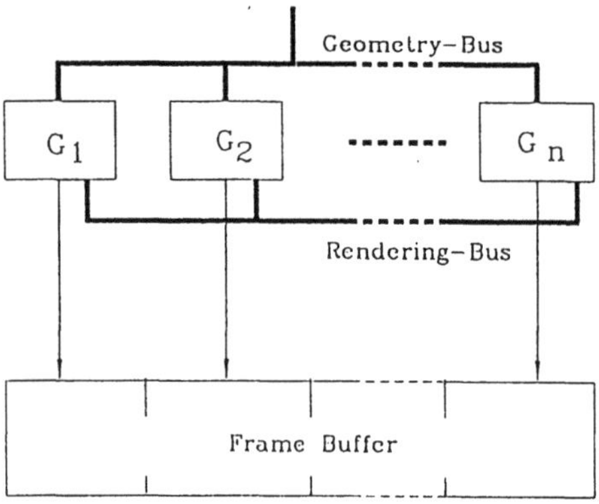

Fig. 5. Structure of the 3DGRP

In order to avoid a bottleneck at the interface between the processors and the memory chips the frame buffer is divided into several banks. Each bank is tightly attached to one processor, which accesses this bank exclusively and need not to compete with others (distributed frame buffer). This produces enhanced pixel data transport to and from the frame buffer required for Bitblt operations and pixel transfer via DMA controllers from the processor memory.

As in Torborg's approach the objects are distributed to the processors GRi via the geometry bus, which is implemented as multiprocessor bus as well as a broadcast bus. If a GR processor fetches a new geometric primitive to be processed the geometry bus acts as a normal multiprocessor bus. Enabling the geometry bus in the broadcast mode it is a means to speed up initialisation phase and to shorten overall transfer time. This is usefull while updating context data within the memory of the GR modules.

Each processor performs the geometry processing for one object completely. The outcoming data are fed to the processors of the system via the rendering bus, which is similar to the geometry bus a broadcast bus as well as a normal multiprocessor bus. Then the processors render the objects for all those pixels, which are situated in the attached frame buffer bank. That means, the geometry processing of one object

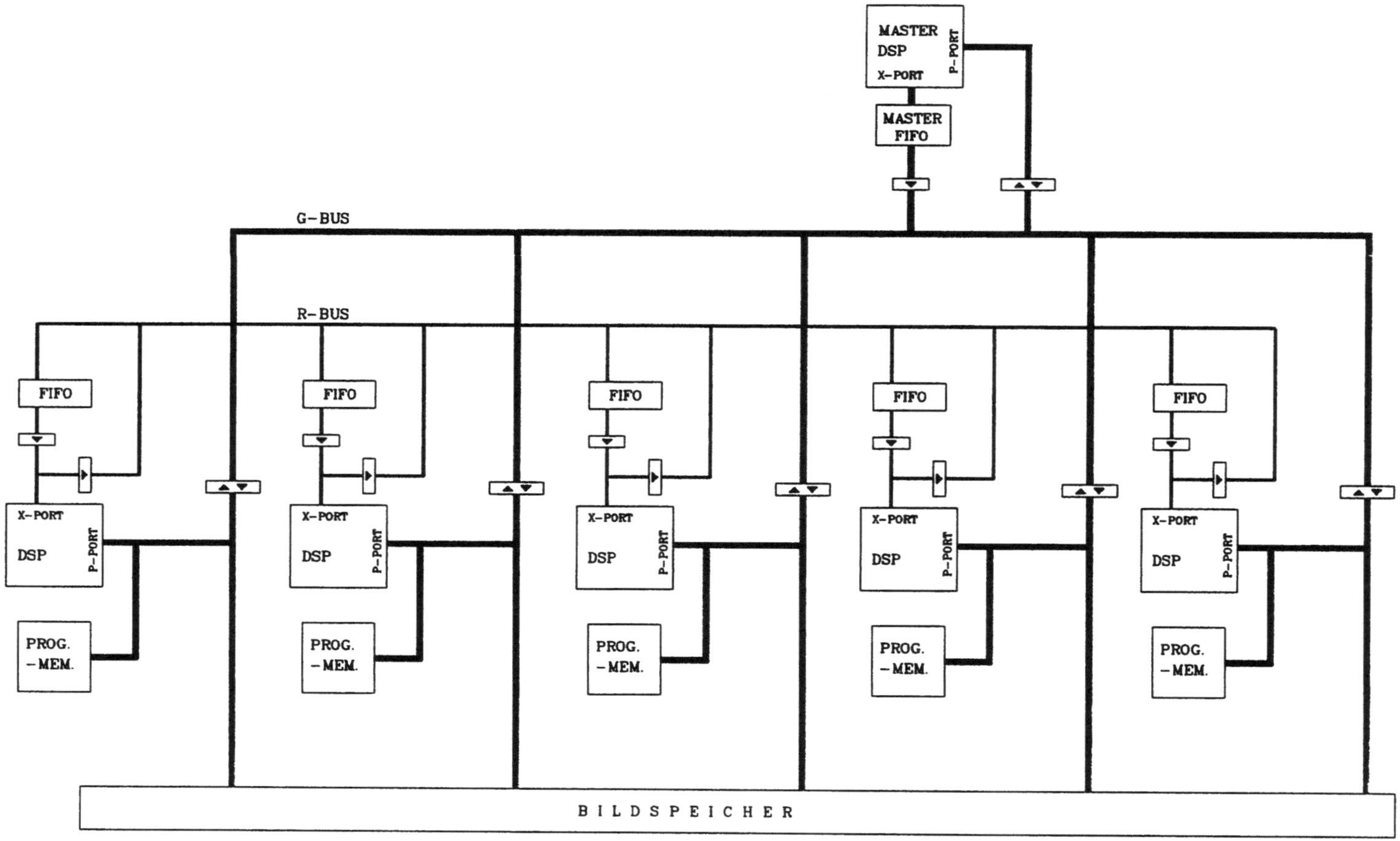

Fig. 6. 3DGRP architecture

is done by only one processor. However, the rendering task of this object is done by all the processors but with reduced pixel amount.

The logical interface between the geometry and rendering calculations transfers pixel, vectors, triangles and trapeziums with edges parallel to the y axis as rendering primitives to the GR modules. Because of being preprocessed e.g. triangle data are calculated really fast. The start values for a span are computed in less than 500 ns and a pixel within a span is delivered every 1 μs. Note that these values are valid for true color, z-buffered Gouraud shading and all the calculations are done with full floating point precision thus reducing computational inaccuracy to a minimum.

The effort of computation for processing the incoming data is dependent on size and position of the geometric primitives. Small triangles or short vectors in parallel to the x or y axis require only a small number of rendering operations. The greater burden of computation relative to the resulting number of pixel of that primitive is in the geometry section. If there are very big triangles much more processing power for rendering calculations is needed than for geometric calculations.

The peak performance of this architecture is limited by the number of processors involved and is delivered when nearly all processing power is exploited for rendering computations. The above mentioned data dependencies cause shifts of processing power requirements between the geometry and the rendering section. This architecture will overcome this difficulty within its realm of performance.

Since any processor may run idle after rendering an object if there is no rendering data in his input buffer it will request new unprocessed objects and continue geometry calculations. In this way an automatic load-balancing is achieved across all the processors. When several GR moduls are doing geometric calculation the overall rendering performance is reduced in favour of geometry processing power but no computational power is going to be wasted by a GR processor starting out to run an idle state.

For this switching mechanism an algorithm is required that changes quickly the context of computation. The internal structure of the processors (single cycle instructions, internal memory) provide a fast task switching within 2 μs or less thus supporting sustained processor performance and avoid the advances of automatic load balancing to be eaten up by switching time.

In Fig. 6 a realization example of 3DRP is given, utilizing five GR-modules. As processor the digital signal processor Texas TMS 320C30, the fastest microprocessor available at present, has been chosen. This processor provides two memory ports, so it can be connected easily to geometry and rendering bus.

The performance of this realization of a 3DGRP architecture is estimated as nearly 400,000 gouraud-shaded 10 pixel vectors per second and nearly 50,000 gouraud-shaded, z-buffered 100 pixel triangles per second.

2.2 Frame Buffer Architecture of 3DGRP

The frame buffer architecture is represented by Fig. 7. The on-screen memory contains all the pixels visible on the screen. The resolution is 1280 × 1024 pixels.

Each pixel consists of 24 bit color data twice (true color, double buffer), 24 bit depth value (Z-buffer), 8 bit transparency, 8 bit window identifier, and 4 bit cursor overlay.

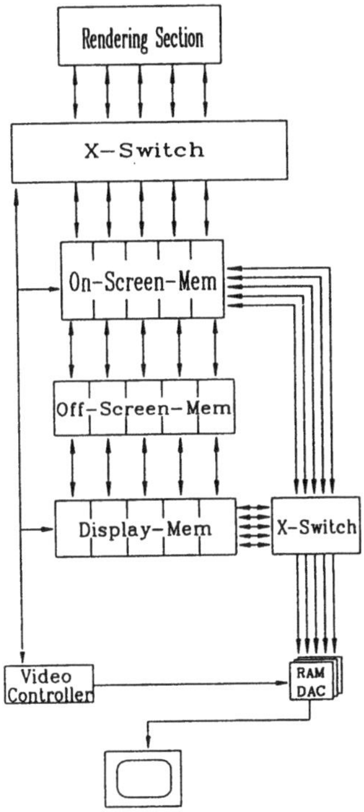

Fig. 7. Frame buffer architecture

The off-screen-memory provides additional memory space for pixel data. It can be used by a window manager for the implementation of a backing store or as a font cache. The off-screen memory contains nearly two million pixels. The use of the display memory is described later on. All of these memories can be accessed by the rendering processors.

As described above the rendering processors are connected to a distributed frame buffer. Three conditions for a reasonable partition of the entire storage have to be taken into account:

- the required display resolution of 1280 × 1024 pixel
- the organization of modern VRAM-chips (256k × 4 bit)
- the number of necessary VRAM-chips, which should be minimal.

This leads to a distributed frame buffer consisting of five equally sized memory banks. Every bank contains 1280 : 5 = 256 columns of the whole screen. The arrangement of these columns is shown in Fig. 8.

In order to realize BitBlT operations within the distributed frame buffer every processor needs to access all the frame buffer banks. Therefore a crossbar switch (X-switch) connects all memory banks with the rendering processors.

Every pixel has a 8 bit window identifier (window-ID) which is used to index a window look up table (WLUT). The window-ID controls write accesses of the

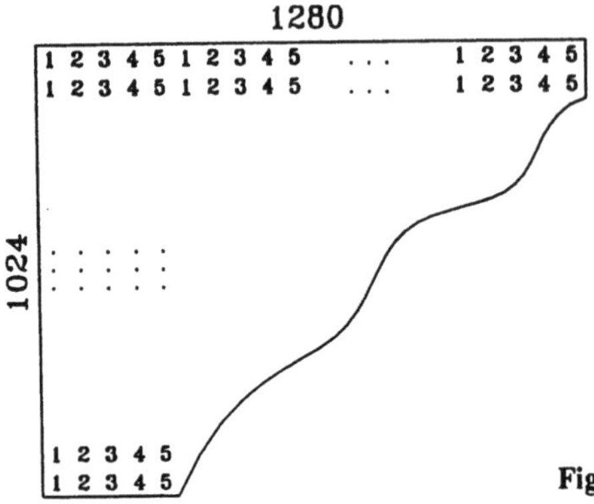

Fig. 8. Frame buffer interleaving

rendering processors to the frame buffer as well as presents basic information for the video controller:

- Controlling the rendering processors' write accesses allows to handle arbitrarily shaped windows. Due to this for every write operation the rendering processors have to compare the window-ID of the new pixel with the stored one. In case of difference the write operation is cancelled. This mechanism is also used for off-screen memory access.
- The video controller utilizes the output of the WLUT to select for every window individually the source of pixel data which are to be displayed on the monitor (double buffering individually for every window).
- The video controller can fill a window of any size and shape with a constant colour during one frame cycle by evaluating the WLUT output.
- The window-ID manages to select one out of five colour look up tables (CLUT) for every window individually.

Finally, advantage can be taken of the window-ID to support very efficiently moving of multiple windows across the screen. Therefore the video controller is able to copy these windows without intervention of the processors. Before starting the copy function the window manager has to provide a shift vector, the identifiers of the windows to move and all identifiers of those windows which are allowed to cover the ones to move. While the copying is done by the video controller the rendering processors can already start to restore the formerly overlapped screen areas. The hardware copy function handles the visible part of the double buffer (color data), the window-IDs and optionally the cursor data.

The serial input/output ports of the VRAM chips of the on-screen and the display memory and the input ports of the RAMDACs are linked via the second crossbar switch. Normally the SAM ports of the on-screen memory are switched directly to the RAMDACs for a conventional screen refresh and to the SAM ports of the display memory to store all the pixel information of the present frame. This is a necessary precondition (Fig. 9a) before starting the hardware supported window copying described by the following example.

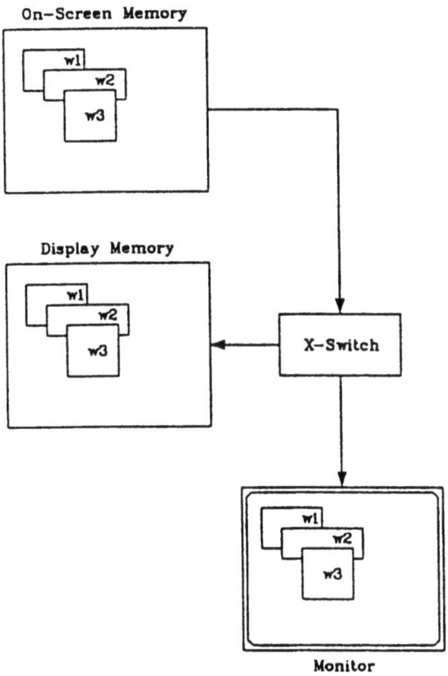

Fig. 9a

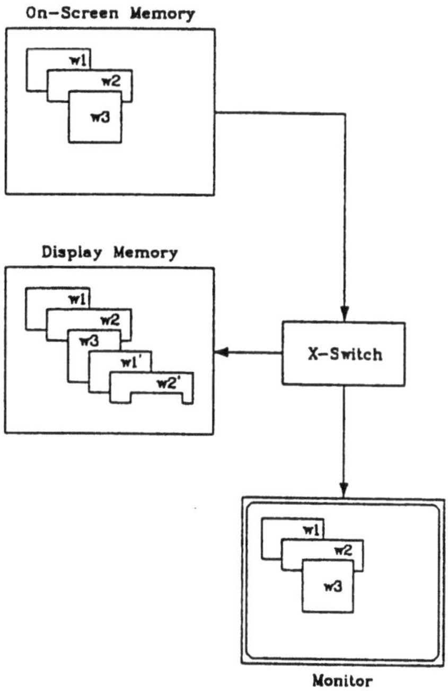

Fig. 9b

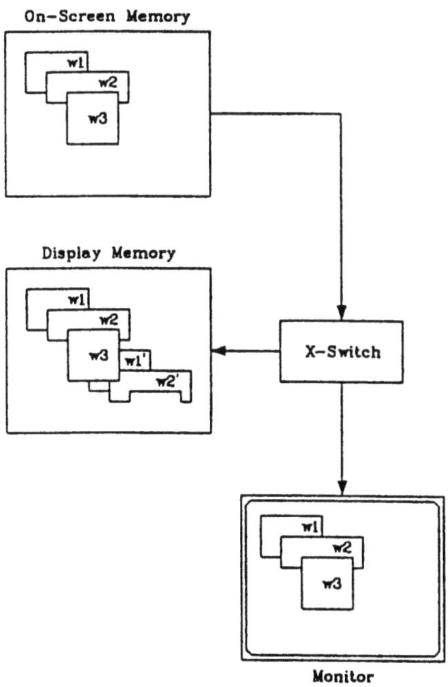

Fig. 9c

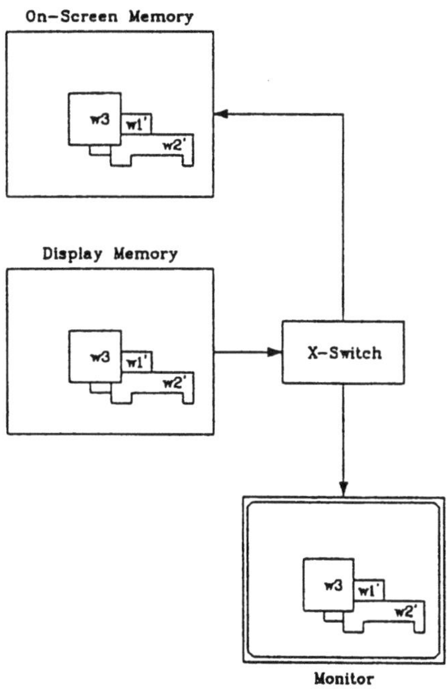

Fig. 9d

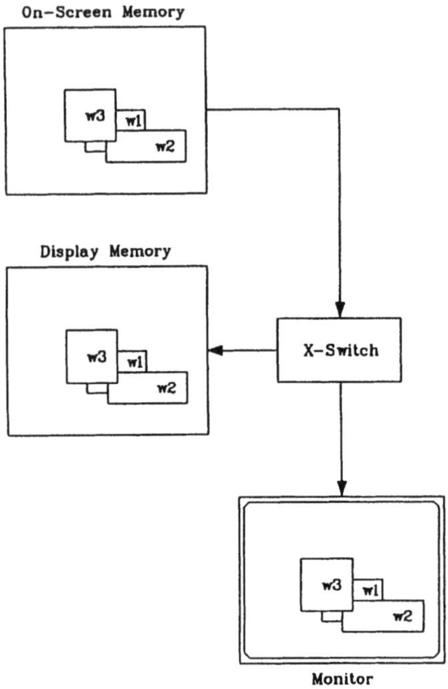

Fig. 9e

A screen image including three windows W1, W2, W3 which is displayed on the monitor is stored in the on-screen and display memory. The joint windows W1, W2 are to be moved from the upper left corner (see Fig. 9b) in direction to the lower right. For this example the window W3 can occlude the windows W1 and W2 and the window W2 covers W1. The mechanism of the hardware supported window copying is illustrated by Fig. 9b to 9e.

The windows to be moved (W1, W2) are copied selectively into the display memory within one frame cycle. Simultaneously the contents of the on-screen memory is displayed on the monitor (Fig. 9b). Therefore the on-screen memory is read as in normal video refresh mode. The serial input stream to the display memory is controlled by two parameters: The window identifier related to every pixel determines (via the WLUT) if a pixel is to be overwritten in the display memory. The second parameter is a starting address of the display memory, which is calculated of a shift vector given by the window manager.

In the following step all windows which may cover the ones to be moved (window W3 for this example) are copied into the display memory without regarding the shift vector (Fig. 9c). That is, the start addresses of the on-screen and display memory are the same. The selective copying is done as described in the previous step and again takes the time of one frame. The monitor input is still generated by the contents of the on-screen memory.

In the next frame, the video refresh is done by the display memory output. Simultaneously all pixels of the display memory are transferred via the serial links to the same location in the on-screen memory (Fig. 9d).

As a result of the copying procedure described above, there may be visible, which have to be regenerated by the rendering processors (e.g. background formerly covered by W1,W2). The rendering processors are free to perform this regeneration during all steps of the copying procedure which are illustrated by Fig. 9b to 9d. The final result is shown in Fig. 9e.

Literature

[AKELY88] Akely, K., Jermoluk, T.: High-performance polygon rendering. Computer Graphics *22*, No. 4 (1988) 239–246

[CLARK82] Clark, J.: The geometry engine, a VLSI geometry system for graphics. Computer Graphics *16*, No. 3 (1982)

[ROG85] Rogers, D. F.: Procedural elements for computer graphics. McGraw-Hill, 1985

[TORBO87] Torborg, J.: A parallel processor architecture for graphics arithmetic operations. Computer Graphics *21*, No. 4 (1987) 197–204

[TUN87] Tunick, D.: Powerful display system revs up image and graphics processing. Electronic Design, July 23 (1987)

An Auditory Display for Exploratory Visualization of Multidimensional Data

S. Smith

Abstract

This paper describes an auditory display being developed for a workstation designed to facilitate the visualization of multidimensional data. To establish the context of this work, the paper first discusses the psychophysical basis for using sound to represent multidimensional data, surveys the related literature, and describes the novel graphical technique employed by the workstation to display multidimensional data visually. Next, it gives an overview of the auditory display and its relationship to the visual display, emphasizing the roles played by the apparent location of sound sources and the repetition of sounds in time. Finally, the paper discusses the current prototype sound system and some of the difficulties to be faced in implementing the proposed facility. This work is being done as part of the Exploratory Visualization (Exvis) project at the University of Lowell.

Introduction

The psychophysical basis for the use of sound as a medium for presenting data is that sound has a number of perceptual properties which correspond in well-understood (though not necessarily simple) ways to its physical properties. For example, pitch corresponds closely to frequency, and loudness corresponds closely to intensity. While there are some interactions between pitch and loudness, these attributes can nonetheless be varied independently over broad ranges with predictable results. Among other sound attributes which can be manipulated independently are attack rate (the rate at which a sound grows to maximum loudness) and decay rate (the rate at which a sound dies away). Attack rate can be varied from nearly instantaneous, like the impact of a hammer striking an anvil, to slow and gradual, like the beginning of a tuba note. Decay rate can be varied in a similar manner.

To present multidimensional data in sound, it is necessary to encode data as the values of properties of sound. One way to do this is to establish a mapping between the dimensions of each data point and the properties of a discrete sonic event, such as a tone or burst of noise. For example, one data dimension could be mapped to pitch, another to attack rate, and a third to loudness. The measure on each dimension of the data point determines the value of the corresponding sound property of the event. The values of all the measures together determine a specific sonic event. In this way a set of data samples can generate a set of sonic events which may be played individually, sequentially, or simultaneously.

The perception of the various properties of sound is described in a vast literature drawn from physiology, psychoacoustics, cognitive psychology, the psychology of music, musical acoustics, and music theory. Roederer [1973] provides a highly readable summary of the pertinent results in these fields. Buxton [1989] offers a thoughtful discussion of the advantages and disadvantages of using specific sound attributes for presenting data.

Related Research

Four studies suggest the potential of sound as a medium for presenting data. Yeung [1980] explored the use of sound as an alternative to graphic presentation of data. In Yeung's experiment, test subjects were asked to classify mineral samples from four sites in California using 7-dimensional chemical data encoded in sound. Each dimension of the data was mapped to a different property of sound and the resulting tones were presented to the subjects. The subjects were able to classify the samples with 90% accuracy without training, and with greater than 98% accuracy after training.

Lunney and Morrison [1981] mapped the high-frequency peaks of infrared spectra onto high-pitched tones, low-frequency peaks onto low-pitched tones, and assigned durations to the tones according to the intensities of the corresponding peaks. The resulting notes were played both sequentially as "melodies" and simultaneously as "chords." In informal tests that required subjects to match tone patterns produced by unknown substances with patterns produced by about a dozen simple organic compounds, subjects rarely failed to match identical patterns.

Bly [1982] investigated the use of sound to add dimensionality to a two-dimensional graphic display. For her experiment, Bly synthesized 6-dimensional data samples and separated them into two sets according to a group of inequalities involving all six variables. Test subjects were to discriminate samples from the two sets when the data were presented visually, aurally, and in a combination of both. All six dimensions were mapped to properties of sound, but only two dimensions were mapped to the x-y scatterplot display. Trained subjects averaged 62% correct identification of samples with graphics alone, 64.5% correct with sound alone, and 69% correct with the combined graphics and sound presentation. Several months after the original experiment, subjects who had participated in the sound-only trials were given additional training. In a second experiment, these subjects correctly identified 74% of the test samples.

Mezrich, Frysinger, and Slivjanovski [1984] used completely redundant visual and auditory displays to present multivariate time series data. Visually, each variable was represented by a pair of vertical lines which got longer and farther apart as the value of the variable increased, shorter and closer together as it decreased. Simultaneously, each variable was represented by a tone whose pitch rose or fell according to the value of the variable. The set of lines and tones corresponding to a point in time was treated as a frame of a movie; the set of frames displayed in sequence at a suitable rate constituted the movie. In the experiment, test subjects had to discriminate correlated and uncorrelated data for four different time-series variables. The data were presented in four different formats: four separate graphs, four graphs aligned vertically, four graphs overlaid on the same pair of axes, and the movie format – four pairs of moving lines with sound accompaniment. The subjects performed this task significantly better with overlaid graphs and the movie technique than with separate graphs and vertically-aligned graphs. For sequences of ten or so frames, the subjects performed significantly better with the movie technique than with overlaid graphs.

Iconographic Display of Data

The critical requirement of an effective data display is that it stimulate spontaneous perceptions of structure in data. The scatterplot, for example, is effective because it stimulates a natural capacity to sense the clustering of points in space. The Exploratory Visualization (Exvis) project employs a novel "iconographic" approach that exploits another spontaneous perceptual capacity: the ability to sense and discriminate texture (see Gibson [1950], Pickett [1970]). The iconographic approach permits a data analyst to use both visual and auditory texture to visualize data.

To stimulate visual texture perception, many small but discriminable elements must be displayed over a relatively small area. Such an image is experienced, for example, when one looks at an expanse of wall-to-wall carpeting. With deliberate visual analysis, one can obtain specific kinds of information about the lengths and materials of individual fibers, but one also receives, without any deliberate effort, an impression of the overall texture of the carpeting. It is just such natural impressions of texture that Exvis seeks to exploit.

In Exvis, each data sample is represented by a small graphical unit called an "icon." An icon's attributes, such as the sizes and positions of its parts, are data driven. Many icons deployed densely over the screen create an "iconographic display." The icons in an iconographic display are analogous to the fibers in a carpet. Just as footprints or the sweep of a vacuum cleaner create variations in the texture of a carpet, so changes in the shape, size, spacing, and orientation of icons create gradients or contours in the visual texture of the iconographic display. The challenge, of course, is to design icons that create highly distinctive textures. The iconographic display technique is described in detail in Grinstein, Pickett, and Williams [1989].

Every icon has auditory, as well as visual, attributes. Depending on the values assigned to each of these attributes, an icon can produce a single tone or other sound on command. Multiple icons sounding simultaneously produce an auditory texture analogous to visual texture. The term "texture" is intended here in more or less its conventional musical sense, that is, the overall quality or surface character of music, which arises from the relationships among its harmonic and melodic components. The auditory textures generated by our system would hardly be mistaken for traditional music, however. Typically these sounds are reminiscent of musique concrete, atonal music, aleatoric music, and other forms of "modern" music.

An analyst may elect to use only vision, only sound, or both. The same fields of the data sample that are used to determine that auditory attributes of an icon may also be used to determine the visual attributes. Alternatively, some or all of the data fields that determine the visual attributes may be different from those that determine the auditory attributes. Thus, the visual and auditory attributes may be used redundantly to reinforce each other or independently to maximize the number of data dimensions the icon can represent.

The Auditory Display

The visualization workstation we are developing will present the user with a rectangular visual display window and a slightly larger concentric rectangular auditory display window. The visual and auditory display spaces may extend beyond the physical visual and auditory windows. When this is the case, the workstation will allow the user to pan over the display space. The workstation will also provide both visual and auditory zoom capability. As the user zooms in on a region of the display, the icons in that region will come acoustically to the foreground at the same time that they grow visibly larger.

Each icon in a display will be able to generate its sound simultaneously with, and independent of, all other icons. By default, each icon's sound will emanate from a direction correlated with the icon's position in the visual display space and in apparent depth relative to the listener. The auditory display will have two "real" dimensions, the horizontal and vertical dimensions of the auditory display window. Depth will be an auditory illusion analogous to perspective in the visual arts: synthesized distance cues will permit each sound to be presented in acoustic perspective "in front of" or "behind" other sounds. Accordingly, the auditory display could be characterized as "2.5 dimensional", with apparent depth not granted full status as a dimension. Because the user will be free to map any desired data fields to the sound parameters which determine the x-y coordinates and depth of the sound source, spatial realism will be a user-selected, not fixed, characteristic of the auditory display.

Each icon will sound periodically at a rate determined by a selected data field. The repetitions of icons having values that fall within a specified range in the selected field will be synchronized, but not necessarily simultaneous (the attacks may be staggered by a fixed time interval). There are two reasons for turning icon sound on and off rather than allowing the icons to sound continuously. The first is that one cannot guarantee that an arbitrary mapping of data fields to sound attributes will produce a pleasing, well-orchestrated sound that reveals structure in the data. The results is far more likely to be cacophony. Turning each icon's sound on and off increases the likelihood that icons with closely similar values in the data field that determines repetition rate will be heard as a group, particularly if the icons are in a localized region of the display space. This effect will be especially important when sound calls the analyst's attention to structures not evident in the visual representation of the data. The second reason for turning icon sound on and off is to prevent psychological adaptation: if sound is continuous, the user ceases to be conscious of it after a relatively short time.

The introduction of icon repetition rate brings the dimension of time into the workstation model explicitly. At least three different time scales will have to be considered:

1. *The sound microstructure level.* This is the time scale on which individual sound oscillations occur, roughly 0.0001 sec. to 0.1 sec. Timbre (a term for the set of sound attributes that enable us to distinguish one kind of sound source or musical instrument from another) is determined at this level. If adequate control can be exercised over changes in timbre, timbre should provide a powerful set of attributes for representing data.
2. *The note level.* This is the time scale on which effects such as vibrato and tremolo, effects that occur during the life of a single note, are found. This time scale is roughly 0.1 sec to a few seconds. Both the rate and depth of vibrato and tremolo are useful as data-driven parameters on this level. The rotation, oscillation, and blinking of visible parts of icons also occur on this level.
3. *The phrase level.* This is the time scale on which groups of notes are organized into meaningful units in conventional music. This time scale runs from one or two seconds upward; thus, it partially overlaps the note level. Icon repetition occurs on the phrase level. Longer-term changes in the visible intensity of icons (e.g., to identify groups of icons related in some way) occurs on this level.

The workstation will allow the user to map data dimensions to sound attributes on all of these levels. This flexibility should allow the user to "tune" the auditory display to reveal structures in the data most vividly.

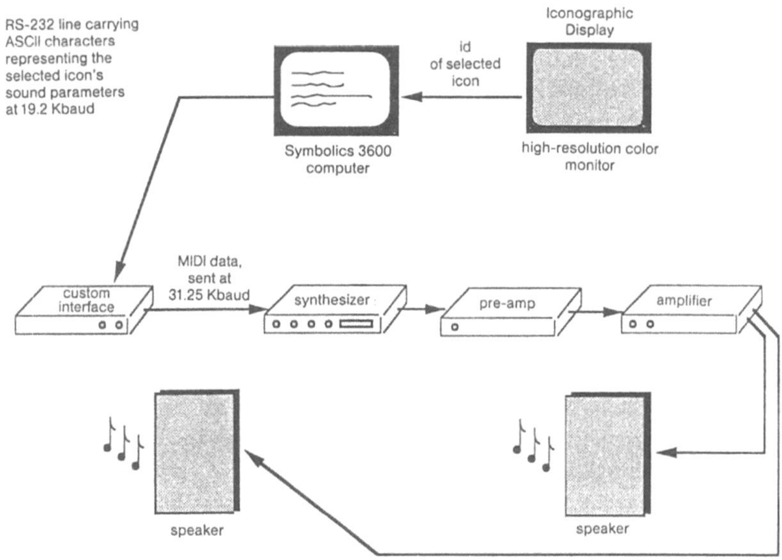

Fig. 1. Schematic of sound production in the current prototype

Workstation Prototype

A workstation prototype has been implemented on a Symbolics 3600. This system is described in Smith and Williams [1989]. Fig. 1 shows the hardware components of the prototype. The sound facility of the prototype workstation associates an independent musical "voice" with each icon of an iconographic display. The large number of logical voices represented by all the icons in a typical iconographic display are realized by eight physical voices belonging to a Yamaha FB-01 tone generator. The physical voices are shared in round-robin fashion by the icons. Each physical voice is capable of producing a time-varying audio waveform under data control. The kinds of sounds that can be generated range from musical tones, to more complex sonorities reminiscent of gongs and bells, to noise.

Each icon's sound is produced when the user moves the cursor into a specified region around the icon. The character of each icon's sound is determined by the values assigned to the parameters of its voice. The values are derived from the fields of the data sample which the icon represents. In the current prototype, the sound parameters available at the level of the individual icon are pitch, loudness, attack rate, decay rate, and depth of frequency modulation (which determines the number and strength of the frequency components in the spectrum of the sound).

The sounds of all the icons within an iconographic display are variations of a single basic sonority. The prototype allows the user to specify global sound attributes for all icons, thereby giving the display as a whole a particular sonic character. The principal global attribute is a tuning ratio that determines the frequency-domain spacing of the spectral components of the sounds generated.

Other important global attributes are vibrato/tremolo rate, vibrato depth, and tremolo depth. We would have preferred these attributes to be individually data-driven for each icon, rather than global; however, the FB-01 tone generator used in the prototype provides only one low-frequency oscillator to control tremolo and vibrato for all eight voices.

If the user sweeps the cursor over many icons, the resulting multiple attacks and overlapping tones can produce auditory textures analogous to the visual textures produced by the icons. These textures are affected by the way the user moves the cursor: fast or slow, in a linear or circular motion, over a small or wide area of the screen, etc. Cursor movement determines, for example, how many notes start per unit of time, how many notes sound simultaneously, and how many different timbres may be sounding at once. In a sense, then, an iconographic display is a two-dimensional musical score which the user can play as he or she wishes.

Using the prototype, we have already accumulated some experience in the use of sound in an iconographic visualization environment. Some preliminary results obtained in experiments by Marian Williams may be found in Williams, Smith, and Pecelli [1989].

Discussion

The prototype sound facility provides a single-icon aperture onto the data. Accordingly, it could be said to have zero spatial dimensions. A one spatial dimension facility would provide a wider aperture (i.e., more icons sounding simultaneously) and the ability to place the apparent source of each icon's sound anywhere horizontally in front of the user. These capabilities, combined with data-driven repetition rate, would permit clusters of icons with similar values to be heard as a group in some region of the left-to-right sound image. Such a facility can be implemented through straightforward adaptation of standard stereophonic recording techniques; moreover, it can be expanded to a "1.5-dimensional" facility by synthesizing distance cues that are present in a reverberant environment, as is routinely done in the production of stereophonic recordings.

Fig. 2 is a schematic of the 1.5-dimensional display we are building as of this writing. The tone generator produces data-driven sounds in much the same way as the tone generator in our current auditory display; however, unlike the tone generator in the current auditory display, this tone generator provides an individual output for each voice. Individual outputs allow the sound for each voice to be processed in a different way, an indispensable capability when each sound is to be given its own unique position in acoustic space. Note that there is a separate mix for each loudspeaker to create the left-right positioning of each sound and a common mix through the reverberation unit to produce the distance cues. The audio level of each voice is controlled at each mixer in real time by MIDI (Musical Instrument Digital Interface) voltage-controlled amplifiers

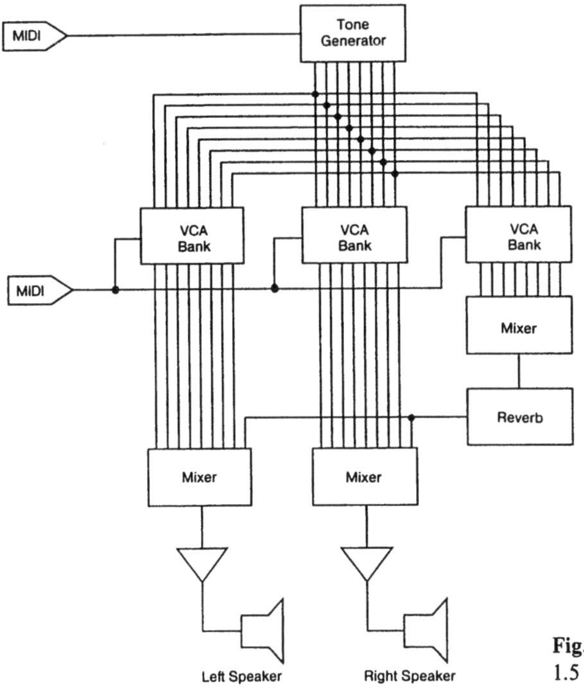

Fig. 2. Schematic of 1.5 Dimensional Auditory Display

(VCA's). Approximately 70 bytes of information are required to specify each sound.

Implementing a two-dimensional sound facility, which allows placement of the apparent sound source anywhere in an imaginary vertical window in front of the user, is more problematic. A two-dimensional array of dozens of small loudspeakers would work but it would be expensive and cumbersome, and it would create a whole new set of problems. For example, if the loudspeaker array were placed behind the visual display, the visual display would cause an acoustic shadow that interferes with the desired directional effects; on the other hand, if the loudspeaker array were placed in front of the visual display, it would obscure the visual display. Ideally, the directional cues could be synthesized for headphones or a small number of loudspeakers.

A key problem in achieving a practical two- or three-dimensional auditory display is synthesizing the cues for the elevation of a sound source. There is as yet no standard technique for synthesizing the cues for sound source elevation. Watkins [1978] simulated the elevation of a noise source by combining a very slightly delayed (160–260 microseconds) echo with the direct signal and presenting the sound via specially-modified headphones. The amount of delay determined the apparent elevation of the sound source. This method requires that the noise source contain frequencies above 4 kHz; without these high frequencies, experimental subjects can not localize the sound source. More recently, Wenzel, Wightman, and

Foster [1988] used the direction-dependent filtering characteristics of the outer ear to synthesize a spatially-realistic three-dimensional auditory display requiring only a pair of ordinary stereo earphones. This method is more powerful than Watkins' in that the apparent source of any kind of sound, not just specific types of noise, can be placed anywhere on an imaginary sphere around the listener. Crystal River Engineering of Groveland, CA offers the "Convolvotron", a two-card set for the IBM PC or PC-AT that synthesizes a 3D auditory display of 1–4 sound sources in almost real time according to this method (there is a delay of 30–40 milliseconds for directional controls and audio to propagate through the system). Kendall, Martens, and Decker [1989] and Moore [1989] have developed elegant models for controlling the spatial location of sound sources. Both, however, require lengthy computations that cannot be accomplished in real time with general-purpose computers.

Because interactivity is one of the essential characteristics of Exvis, the auditory display must work in real time. This requirement has led us to use audio equipment based on the MIDI standard. Because MIDI equipment is intended primarily for musical performance, it is designed for real time operation. However, because MIDI originated as a communication protocol for *keyboard* control of digital music synthesizers in the context of popular music, it has inherent limitations that are difficult to work around in an application like ours. Among these limitations are low-bandwidth data communications (31.25 Kbaud), very limited control over the microstructure of sound, and a tendency to become clogged with messages. Because of these limitations, it is likely that the next generation Exvis workstation will use special-purpose hardware for sound generation.

Summary

Experimentation with sound as a medium for representing data has been underway since at least the beginning of the 1980's. The Exvis project brings several new ideas to this work, including the interactive manipulation of sound, iconographic data representation incorporating sound attributes, and the integration of auditory and visual displays into a single, unified visualization facility. Our experience in working with sound on a prototype workstation for scientific data visualization has encouraged us to further develop and refine the auditory display. Our immediate goals are to implement the 1.5D display described here and to determine if the kind of spatial information it provides is useful in the visualization process. If our intuition that such information is useful proves to be correct, we plan to undertake the design and implementation of a 2D or 3D auditory display.

References

Bly, S. A. "Presenting Information in Sound." *Proceedings of the CHI'82 Conference on Human Factors in Computer Systems*, pp. 371–375, 1982

Buxton, W. "The Use of Non-Speech Audio at the Interface." *Tutorial # 10, CHI '89*, pp. 2.1–2.15, 1989

Gibson, J. J. *The Perception of the Visual World.* Boston: Houghton-Mifflin, 1950

Grinstein, Georges, Ronald Pickett, and Marian G. Williams. "EXVIS: An Exploratory Visualization Environment." *Graphics Interface '89.* London, Ontario, 1989

Kendall, Gary S., William L. Martens, and Shawn L. Decker. "Spatial Reverberation: Discussion and Demonstration." In M. V. Mathews and J. R. Pierce, *Current Directions in Computer Music Research.* Cambridge: MIT Press, 1989

Lunney, D. and R. C. Morrison. "High technology laboratory aids for visually handicapped chemistry students." *Journal of Chemical Education* 58(3):228–231, 1981

Mezrich, J. J., S. Frysinger, and R. Slivjanovski. "Dynamic Representation of Multivariate Time Series Data." *Journal of the American Statistical Association* 79(385):34–40, 1984

Moore, F. R. "Spatialization of Sounds over Loudspeakers." In M. V. Mathews and J. R. Pierce, *Current Directions in Computer Music Research.* Cambridge: MIT Press, 1989

Pickett, R. M. "Visual Analyses of Texture in the Detection and Recognition of Objects." In B. S. Lipkin and A. Rosenfeld, *Picture Processing and Psycho-Pictorics.* New York: Academic Press, 1970

Roederer, Juan. *Introduction to the Physics and Psychophysics of Music.* New York: Springer-Verlag, 1973

Smith, Stuart and Marian G. Williams. "The Use of Sound in an Exploratory Visualization Environment." *University of Lowell Computer Science Department Technical Report No. R-89-002*, 1989

Watkins, A. J. "Psychoacoustical aspects of synthesized vertical locale cues." *Journal of the Acoustical Society of America* 63:1152–1165, 1978

Wenzel, Elizabeth M., Frederic L. Wightman, and Scott H. Foster. "Development of a three-dimensional auditory display system." *SIGCHI Bulletin* 20(2):52–57, 1988

Williams, Marian G., Stuart Smith, and Giampiero Pecelli. "Computer-Human Interaction Issues in the Design of an Intelligent Workstation for Scientific Data Visualization (Phase 1)." *ACM SIGCHI Bulletin*, to appear. (Preprints available as University of Lowell Computer Science Department Technical Report No. R-89-006.)

Yeung, E. S. "Pattern Recognition by Audio Representation of Multivariate Analytical Data." *Analytical Chemistry* 52(7):1120–1123, 1980

A Model for the Use of Different Sound Description Layers Within a Multimedia Environment

C. Blum

1. Abstract

The number of applications on graphics workstations using sound in order to enhance the human-computer interaction capabilities is increasing. Using and handling sound in addition to text and graphics seems to be the next forthcoming step towards a multimedia environment.

This paper outlines a first approach for an audio content architecture consisting of different description layers which are able to handle different levels of abstraction. Therefore a comparison between image rendering and sound synthesis is made. Then a short overview on digital sound synthesis techniques and sound color models is given in order to show hardware requirements as well as psychophysical difficulties. Finally, the audio layer model is presented. It contains three layers for parametric and symbolic audio description, one layer for digital audio, and one presentation layer.

2. Introduction

It is evident that presenting information for the two main human sensors, eyes and ears, enlarges the bandwidth for human-computer interaction. The following list shows some typical application areas of graphics workstations dealing with auditory information.

User Interfaces

Since sound plays an integral role for human communication as well as for human orientation in the world, it should be available in every comfortable user interface. Thus, the human-computer dialogue becomes more "natural" and convenient. A lot of everyday information is perceived via auditory cues. When using sound within user interfaces, a lot of information about the internal system state (or state changes), previously hidden or presented visually, can now be presented via auditory cues. An extensive collection of ideas how to use sound at the interface can be found in [Buxt89].

Scientific Visualization

Some research activities have been done in previous years to examine the use of sound for data presentation within scientific visualization environments. The so-called ExVis-prototype, described in [Grin89] and [Smith89], gives an example how to use sound in addition to vision for the visualization of multidimensional data. Sound enhances the visual presentation of data in two ways: it can reinforce data being presented simultaneously through visual attributes and it can be used to increase the number of data dimensions presented at once.

Music

Talking about multimedia, the presentation, storage, processing and synchronization of music with other types of information such as motion picture is one topic to be considered. For example, computer-animated films gain a lot of effects when they are presented in combination with music.

Speech

Speech is certainly the most important form of auditory communication. Although a lot of research activities have been done in the field of speech recognition, it is not yet a commonly available technology. In order to show the great special position of speech, all other applications listed above are subsumed under the term "non-speech-audio".

There are also some applications where speech can be processed without using recognition algorithms: an electronic mail system could handle speech, coded as digital audio sequences, instead of text.

3. A Comparison Between Image Rendering and Sound Synthesis

The left part of figure 1 illustrates the rendering and visualization pipeline. A symbolic or generic description of objects to be visualized is transformed into a parametric description. This step is called visualization and contains a certain data interpretation. The parametric description contains information about object geometry and texture attributes. Together with some environmental variables (e.g. illumination models) it determines the result of the next step called image rendering. Digital images are time- and space-discrete representations. They can easily be translated into perceptable (analog) signals. Thus, the description is called "display-oriented".

Looking at the right part of figure 1, a similar pipeline is presented. Musical scores as well as any other multidimensional data need to be interpreted in order to obtain parametric sound descriptions. The translation of these parametric data

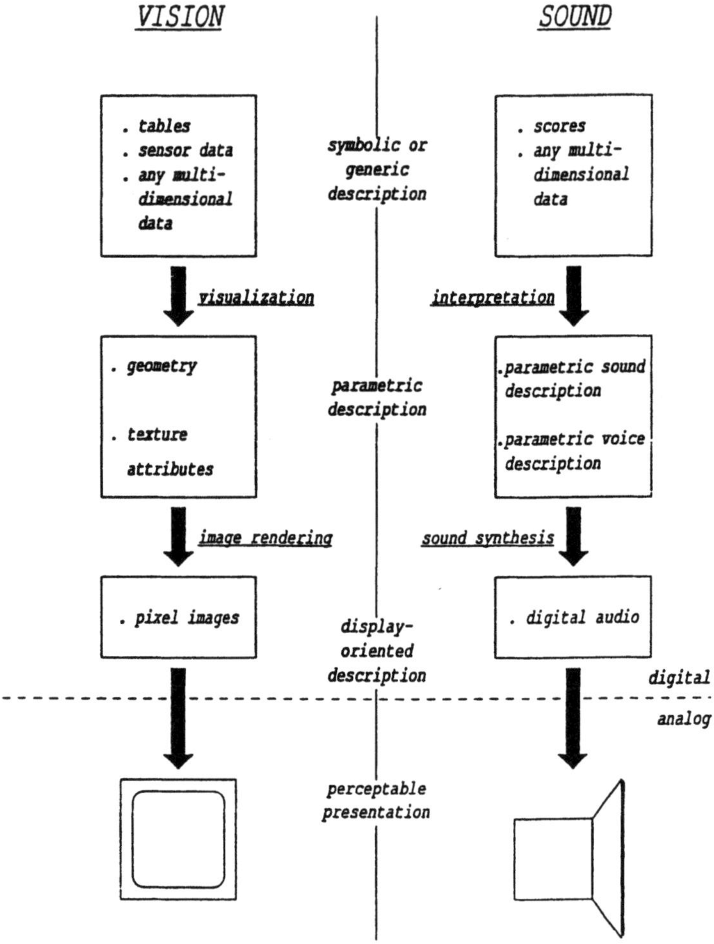

Fig. 1. A comparison between image rendering and sound synthesis

into discrete and display-oriented sound data is called sound synthesis. The result is a sequence of digital sound samples, called digital audio. It is obvious that both the vision and the sound pipeline contain principally similar algorithms and description levels.

Nevertheless, we have to consider some important differences between the two media:

Data Rates and Technology

Let us take a look at the data rates required to present visual and auditory information in state-of-the-art technology (without using any data compression method). Here are some examples:

- a still image presented on a high resolution monitor with 1k * 1k pixel and three 8-bit color channels has a dataset of 24 Mbit,
- a digital motion picture, as it is specified in the CCIR 601 standard, has a data rate of 216 Mbit/s.,
- a high-definition television standard has a data rate of more than 1 Gbit/s., and
- a digital audio sequence with a sampling frequency of 40,000 samples per second and a sample depth of 16 bits per channel has a data rate of 1.28 Mbit per second.

Taking into account that a stereo audio signal of about 1 Mbit per second produces a sound quality which leaves nothing to be desired, it turns out that a much higher data rate (two to three of this order) is required to produce motion pictures even with less satisfying results.

Visual and Auditory Sensors

A human being gets his colorimetric information from three different types of retinal receptors. Each of these receptors integrates over the visual spectrum using a certain weighting function. Consequently the colorimetric space becomes three-dimensional and one cannot distinguish between two different spectra if their resulting receptor values are identical. This phenomenon is called "metamerism". As a result it is sufficient to present a visual spectrum which only consists of three single peaks (usually red, green and blue) instead of presenting the whole original spectrum. For the human's auditory sensors, the situation is different. It is necessary to present the whole audible spectrum. Thus, it is impossible to reduce the information to only three dimensions. The result is either a representation which divides the spectrum into a large number of spectral components or a completely different representation. Examples for both methods can be found below in the discussion of synthesis techniques.

The Role of the Dimension "Time"

The dimension "time" occurs in both the vision pipeline and the auditory pipeline. We look at the world and listen to the world at the same time. Nevertheless, the role of time is different. In [Gave89] an explanation for the differences between vision and sound is given as follows:

> "Sound and vision are complementary modes of information. Sound exists in time and over space, vision exists in space and over time. Sound is available for a limited time but most visual objects, on the other hand, tend to persist. Visual objects exist in space, thus it is necessary to look in the appropriate direction but sound sources need not to be faced."

Looking at physical properties, there is no clear difference between the role of time for sound and its role for vision. Time is one of the four dimensions of our space-time-continuum which is inherent in vision as well as in sound. Both vision and sound are sensations based on a spectrum of frequencies. Only the order of magnitude and the physical medium (electro-magnetic energy vs. sound pressure) are different.

The way we process visual and auditory information makes the big difference. In contrast to vision, certain time-variant components of audio signals are translated into sound color sensations. They are perceived without hearing changes in time. Furthermore, sounds have a beginning and an end, and most sounds are brief enough that both are experienced.

The time based dimensions of sound can be divided corresponding to their timescales (see [Roed73]) for details). Each timescale can cover different types of auditory (or musical) phenomena:

- frequency and waveform belong to the microscopic timescale,
- envelope parameters (e.g. attack time or decay time) and also tremolo belong to the intermediate timescale, and
- duration and all parameters dealing with rhythm belong to the macroscopic timescale.

The importance of time for the perception of sound implies that there is a need for realtime synthesis and realtime processing.

4. Digital Sound Synthesis Techniques and Sound Models

As a matter of fact, sound models correspond to synthesis techniques: a sound synthesis technique is essentially a model (often expressed as a formula) for generating sound. A model for the parametric description of sound is useless if there is no synthesis technique accepting these parameters as an input. On the other hand a sound model should meet some other requirements. It is desired to obtain representations of sound that suggest ways to provide both a sound space being as complete and homogeneous as possible, and low-dimensional control over their perceptually important properties. But what are those properties?

Some studies have been made in order to examine the correspondence between perceptual and physical properties. In most of these studies the perceptual properties are oriented by musical listening having attributes like pitch or timbre (see [Roed73] for an introduction into the psychophysics of musical listening). Sensations such as pitch, loudness or spatial direction (do not forget that sound exists in space) can easily be mapped onto physical properties such as frequency and intensity, in consideration of some restraints. But timbre (or sound color) is clearly multidimensional and its dimensions are only partly known.

Spectral energy distribution and the temporal variation of this distribution provide the acoustical determinants of our perception of sound color. See [Grey75] for a three-dimensional timbre model.

In order to find sound models for the parametric sound description layer of the audio layer model, the "musical" approach is not sufficient by itself. For example, sound events in user interfaces should carry information similar to those carried by so-called everyday sounds, e.g. the sound appearing when two materials collide. In [Gave89] a set of attributes of everyday listening is presented, including attributes such as "material", "size" and "interaction". It is based on the understanding, that in most cases we do not hear sound but instead sources of sound. The more abstract those sound attributes are, the more complicated it is, of course, to map them onto physical properties. Within the audio layer model, all different kinds of sound representations should be supported. There is no single representation which can cover all features.

A lot of work has been done in the field of hardware design for digital sound synthesis. A comprehensive technical overview is given in [Stra85]. There is a large market for (low-cost-) synthesis hardware, mainly designed for musical purposes. These versions are not sufficient for the scientific use. For example, applications in scientific visualization require a more flexible handling of pitch or a larger number of sound generators which are simultaneously available. Rules for the design of synthesis hardware are:

- to minimize the amount and complexity of hardware,
- to maximize the performance (e.g. number of generators or delay time),
- to obtain a maximum variety of sounds,
- to obtain some excellent (mostly "natural") sounds, or
- to obtain input parameters that allow an easy handling.

The following list briefly describes the essential synthesis techniques:

Additive Sound Synthesis

In the additive model for sound synthesis, a tone is represented by the sum of sinusoidal components, each of which has time-varying amplitude and frequency ([Wess85]). The character of the temporal evolution of the spectral components is of critical importance in the determination of timbre. Additive sound synthesis is a general method being able to generate a very large and highly varied class of signals. Historically, additive synthesis is the most important and influential synthesis technique. An important drawback is the tremendous amount of data required to describe all amplitude and frequency envelopes. This may exceed the bandwidth between computer and synthesizer.

Frequency Modulation

In Frequency Modulation (FM), the instantaneous frequency of a carrier wave is varied according to a modulation wave, so that the rate at which the carrier varies is the frequency of the modulating wave, or modulating frequency (see [Chow3]). FM provides a simple temporal control over the bandwidth of spectra whose components can have a variety of relationships. FM turns out to be more economical in terms of the hardware required to produce a given sound and the amount of data needed to specify a sound.

Subtractive Synthesis

In opposite to the additive sound synthesis, the subtractive approach starts with an easily generated signal being rich in spectral components (such as noise or pulse train), and attempts to produce the desired result through a series of filtering operations.

Sampling and Wavetable Lookup Synthesizers

The basic sound characteristic of a sound generated by a sampling synthesizer comes from a short sequence of digital audio. This sample sequence is transformed in realtime in order to get the desired output sound. The transformation is controlled by parameters like frequency, velocity, duration, attack time and decay time. This is a very easy way to generate (or better: reproduce) "natural" sounds (e.g. sounds of natural instruments). The technique can be compared to texture mapping as a part of the visual rendering pipeline. It can produce a high level of realism but its control parameters are limited.

5. The Audio Layer Model

As you can see in figure 2, the architecture consists of four different audio description layers and one physical presentation layer.

The parametric audio description is divided into three different layers in order to provide an interface which is application-independent. The *application layer* contains data and algorithms for a large variety of applications dealing with sound such as scores for musical notation or multidimensional databases and associated mapping algorithms for scientific visualization. Also algorithms for the processing of sound events within user interfaces are included.

All application data are translated into a *uniform parametric sound description*. This description has to be very extensive in order not to be a bottleneck for the variety of applications and the variety of existing synthesis techniques. A list of relevant sound description models, incomplete, however, was gathered in section 4. To have such a uniform description layer allows changes in the system hardware (and the sound synthesis hardware as well) to be made easily without rewriting parts of the application software. Thus, the software becomes portable and auditory results become reproducible. This technique has been applied in the field of computer graphics since the late 70's.

The *synthesis of parametric sound data* can be done either using synthesis *software* running on the host computer (this software should be available as system tools like illumination or shading algorithms for visual rendering), or using *special*

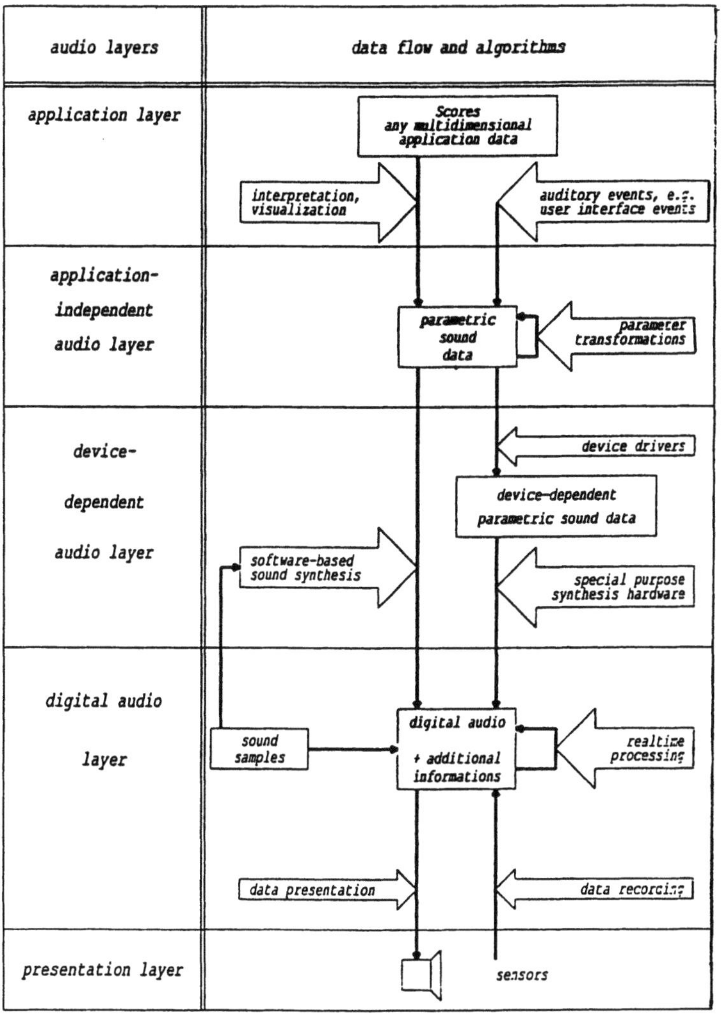

Fig. 2. The audio layer model

purpose synthesis hardware. A *device driver* is required for every synthesizer in order to translate parametric sound data into device-dependent formats.

As we have seen in sections 3 and 4, sound synthesis is based on complex algorithms, and most applications require realtime synthesis. For that reason, it is necessary to have external synthesis hardware. In figure 2 one can see that the output of this external hardware does not reach to the presentation layer (uncontrolled, as it does in most existing systems). In this model the output is linked to the digital audio layer. Thus, it is possible to manipulate and process digital audio data independently from the synthesis technique used to create these data.

Standard functions required for the *processing of digital audio data* are: cutting, cueing, playing forward and backward, and so forth. Also storage and retrieval operations are required.

A *database* containing short audio sample sequences is part of the audio layer model. This database can be applied for producing "natural" sounds in realtime using software synthesis technique. Most of the algorithms at this level of the audio layer model have to run in *realtime*. For the transfer of audio data in realtime, data rates of about 1 Mbit per second are required.

The opposite of sound synthesis, *sound analysis*, is not presented in this picture with the exeption of digital audio recording (using input sensors such as microphones); it would go beyond the scope of this paper. Nevertheless, it is important to notice at least one important application area for sound analysis: speech recognition. This is getting very important for the next generation of human-computer interaction methods.

6. Standards and Standardization Activities

In order to provide frictionless communication and data exchange between different audio processing workstations, it is necessary to consider certain standards and standardization activities.

Some standards in the area of digital audio coding have been worked out by using different methods of redundancy reduction and different bandwidths. Concerning high-level parametric audio description, there is still a lack of proper standards. Of course, there is the well known MIDI standard, but as you see below, MIDI does not meet all requirements.

Musical Instrument Digital Interface (MIDI)

The MIDI format can be described as a musical script. It covers all information produced when playing on a keyboard: *what note is played at what time and with what velocity*. There are additional parameters like pitch change (for detuning) or control change (for any sound modulation), and some other control parameters for synchronizing several MIDI devices. Using MIDI to control external signal processing equipment allows designers to overcome the signal processing limitations of the host computer, but also bears some other limitations or drawbacks.

MIDI is designed for musical purposes. Frequencies are described in the music-oriented halftone grid, data rates are determined by requirements for musical purposes. A very important limitation is the mechanism used to handle sound colors. There is no way to define sound colors but to set an index on a device-dependent sound color table. The MIDI standard provides a possibility to send so-called "system exclusive messages" in order to control system parameters, but those parameters are highly hardware-dependent. They are different even for synthesizers using the same synthesis principle. Using those control parameters to describe sound means to leave the standard.

Thus, strictly speaking, the MIDI standard format does not deal with sound. It is no standard for parametric sound description. In terms of the audio layer model it provides a symbolic description, useful for the "musical" subset of general audio (e.g. score description).

The standard also covers a physical interconnection specification which has also its drawbacks. See [Loy85] for a detailed discussion whether MIDI is a bus or not.

Standards for Digital Audio Coding

There already exist several standards for digital audio coding. The following parameters and methods determine the quality, the information content and the "functionality" of a sequence of digital audio samples:

- method of data acquisition (recording or generation):
  filtering techniques, bandwidth of the source signal
- data format:
  sampling frequency, sample depth, number of channels
- coding:
  data structures (structured for storage, retrieval or editing operations) and data compression methods (redundancy reduction, irrelevancy reduction)

Here is a brief listing of standardization activities for digital audio:

- The Audio Engineering Society has recommended 48 kHz and 44.1 kHz for sampling rate standards.
- ISO/IEC JTC1/SC18/WG5 has identified the following existing coding standards as possibly useful (see [ISO88] for more information):
  - 64 Kb PCMD (CCITT Rec. G.711)
  - 32 Kb ADPCM (CCITT Rec. G.721)
  - 64 Kb Wide band speech coding (CCITT Rec. G.722)
  - Digital Cirquit Multiplication (CCITT draft Rec. G.72z)
  - 16 Kb Linear Predictive Coding (under development by CCITT)
- As an extension to the MIDI standard format, a so-called MIDI sample dump standard has been developed for the exchange of sampling data on MIDI networks.
- The technique of irrelevancy reduction for audio signals is based on the knowledge of the psychophysical nature of human hearing. See [Stol88] and [Gerh89] for irrelevancy reduction methods, dividing the auditory bandwidth into sub-bands in order to achieve a frequency-dependent quantization.

7. Conclusion

The audio layer model presented in section 5 is a first approach to a general audio content architecture. This paper has outlined that there is, on the one hand, an increasing number of applications on graphics workstations processing audio data, and on the other hand there is a lack of standardized concepts and satisfying techniques. Efforts which have already been made at the vision pipeline have to be effected at the auditory pipeline in the future, keeping in mind that some characteristics (e.g. the role of time) are different in both systems. The intention is to design an open system for general purposes, whereas audio is one type of information among others belonging to a multimedia environment.

Further work will be defining data structures for the different sound description layers mentioned above and implementing tools to handle, manipulate, and synchronize sound information together with other information types such as motion pictures (e.g. on a realtime disk). A lot of work has to be done in order to clearly specify the application-independent audio layer for parametric sound description. A lot of different parameter spaces have to be united in order to support a large spectrum of applications. Finally, the inverse of the generic pipeline described in figure 1 and 2, the sound analysis and particularly the speech analysis, have to be integrated in the model.

Within the cooperation between the University of Lowell and the Darmstadt Computer Graphics Center, future implementations will be based on the ExVis-software[1] which provides the exploratory visualization of multidimensional data using both sound and vision (see [Grin89] and [Smith89]).

8. Acknowledgments

In parts, this work belongs to the BERKOM research project "Multimedia Documents in Integrated Broadband Environments (B-ISDN)", funded by the Deutsche Telepost Consulting GmbH, Bonn (DETECON).

The author appreciates the access to the ExVis-software as well as all discussions during his stay at the Graphics Research Laboratory, University of Lowell, MA.

---

[1] The ExVis-software has been developed at the Graphics Research Laboratory, University of Lowell, MA. and has been installed at the Fraunhofer Computer Graphics Research Group (FhG-AGD), Darmstadt, West-Germany.

References

[Buxt89] William Buxton, William Gaver and Sara Bly: Tutorial Notes of the 1989 SIGCHI conference on "The Use of Non-Speech Audio at the Interface", ACM 1989

[Chow73] John M. Chowing: "The Synthesis of Complex Audio Spectra by Means of Frequency Modulation", Journal of the Audio Engineering Society, Vol. 21(7), 1973

[Gave89] William W. Gaver: "The SonicFinder", Tutorial Notes of the 1989 SIGCHI conference on "The Use of Non-Speech Audio at the Interface", ACM 1989

[Gerh89] Heinz Gerhäuser und Dieter Seitzer: "Musikcodierung - Signalverarbeitung in Echtzeit auf psychoakustischer Grundlage", FhG-Bericht 1/89

[Grey75] J. M. Grey: "An Exploration of Musical Timbre", PhD. diss., Dept. of Psychology, Stanford University, 1975

[Grin89] Georges G. Grinstein, Ronald M. Picket, Marian G. Williams: "EXVIS: An Exploratory Visualization Environment", Graphics Research Laboratory, University of Lowell, 1989

[ISO88] ISO/IEC JTC1/SC18/WG5 document N638: "Liaison Statement on User Requirements for an Audio Content Architecture", 1988

[Loy85] Gareth Loy: "Musicians make a Standard: The MIDI Phenomenon", Computer Music Journal, Vol. 9, No. 4, 1985

[Roed73] Juan G. Roeder: "Introduction to the Physics and Psychophysics of Music", Springer Verlag, New York - Heidelberg - Berlin, 1973

[Smith89] Stuart Smith: "An Auditory Display for Exploratory Visualization of Multidimensional Data", Computer Science Department, University of Lowell, 1989

[Stra85] John Strawn and Curtis Roads: "Synthesizer Hardware and Engineering", chapter 2 of "Foundations of Computer Music", edited by John Strawn and Curis Roads, MIT Press, Cambridge, MA., 1987

[Stol88] G. Stoll und G. Theile: "MASCAM: Minimale Datenrate durch Berücksichtigung der Gehöreigenschaften bei der Codierung hochwertiger Tonsignale", Fernseh- und Kino-Technik Nr. 11/1988

[Wess85] David L. Wessel: "Timbre Space as a Musical Structure", chapter 5 of "Foundations of Computer Music", edited by John Strawn and Curtis Roads, MIT Press, Cambridge, MA., 1987

4. Experimental Workstations and Environments

Some Thoughts on a Computer Workstation for the Biological Sciences

T. O. McCracken, T. L. Spurgeon, L. G. Lodise, S. K. Koch, R. Miranda, and C. Fedde

The biological sciences are in need of a mechanism to consolidate vast amounts of morphologic information. Currently, both academic and professional communities are struggling with the organization and accessibility of current technologies. The computer workstation provides a powerful solution to this management problem because of its ability to adaptively store, access, and communicate information. The computer can adapt to the progression of knowledge, helping the user to assimilate and easily access current technology. The effective use of a computer for this purpose will require complex biological models derived from all the applicable sciences, including anatomy, physiology, biochemistry, kinesiology, and biomechanics.

The biological sciences workstation of the (not-too-distant we hope!) future will be a complete information management tool for the working professional. We see the following broad features as critical to its performance. It will have access to large and varied data bases, providing a wide range of information on biological structures, their anatomy and physiology. Though not limited to this, the primary focus of the workstation will be graphics display, incorporating state-of-the-art imaging technology, and using both surface and volume rendering techniques as the data source and particular application demand. The display features should provide for an understanding of interrelated anatomical structures through an array of graphical tools, including the layering of structures, accurate color and texture representation, and motion, both of internally static and of dynamic elements. Overall, the workstation should house an intelligent expert system which would maximize the flexibility of the user's access to the desired data. On the basis of high-level commands from the user, the system should be able to choose the most appropriate response, including judgments as to data representation, level of detail, etc. Moreover, it should respond interactively to the user's needs and demands in real time. Finally, it will incorporate artificial intelligence to learn from its previous experience to better serve the users of the future.

Large amounts and diverse forms of information would allow the level of representation to be extremely flexible. Information would be available to view a structure at the gross anatomic level, the cellular level, and even down to the chemical/molecular level. The ability to modify conditions and view their effects would be most advantageous. Not only could the practitioner analyze the efficacy of different treatment techniques, but those who wish to view the cause and effect of an injury could set up conditions and replay different aspects of an accident. Imagine the impact on the medical legal profession to actually "see" a case take place.

But quantity alone will not provide quality. The user can easily be overwhelmed with excessive information. In fact, the user shouldn't always be bothered with deciding what view is best to receive information. We envision a machine that, from its vast knowledge of anatomical structure, will analyze the most appropriate approach and amount of detail for the viewer's capabilities and interests. For instance, one wishing to view a gross structure may not need to see every aspect of the surrounding tissues, but perhaps only references to their presence.

Anatomy is understood through a progression of layers (i.e. under the skin lies muscle, under which is the peritoneum, under which lies bone or organs with their complex innervation and vascular supply, etc.). Ideally, the user should be able to "peel away" superficial layers to expose and emphasize underlying layers. This could be accomplished through variable degrees of transparency and/or the progressive simplification of each examined layer. This implies that complete rendering of overlying structures can be an intrusive addition to the image, even though it would be accurate information. It might be more successful to use a surface outline of the skin (or a semitransparent representation) instead of a fully rendered version when the interest lies in underlying organs. However, the layering reference is essential to the understanding of an organ's role in the living organism.

The traditionally layered approach to anatomy also suggests that visual presentation can be partially provided by surface modeling techniques (i.e. polygonal tiling of surfaces), which are currently predominant in computer graphics. However, biological structures also occupy space and have specific space-filling properties, including tissue makeup, texture, density, color, size, and shape. This suggests that data structures and graphical display may more appropriately be based on a spatial occupancy scheme. With an acceptable degree of resolution, a volumetric model can represent surfaces as well as space-filling properties.

Motion is the key to effective computer imaging. Often an image is not completely understood as occupying three-dimensional space until it is rotated and other views are made available. Stereovision and parallax are useful, but dynamic motion of the image is of greatest importance. Not only motion of a static object but the accurate kinetics inherent to the given structures will increase spatial and physiological understanding by revealing the many interactions that take place hidden beneath the skin. With an advanced understanding of biomechanics, the computer would have the ability to analyze possible motions of any given joint, organ, or fluid flow. For example, given the structural data of an elbow joint, the computer would model bone, muscle, and tendon connections, simulate muscle contraction (using aspects of strength from the number of muscle fibers present and the degree of fine motion by the innervation density), and ultimately show the joint in motion. Limits of motion would be provided by the structural inhibitions of the joint, just as they are in the living organism. However, it would be interesting to be able to overcome these natural limits and view the effects of such an action.

One of the primary functions of biomedical imaging is to promote understanding by simplification. The computer must be able to make this simplification, but with flexibility. If the viewer is still not able to understand the image, or is unable to find answers to inquiries, the computer should be able to change its approach through the use of analogies, schematics, and reference materials. If more detail is requested, it should be quickly provided. Should the user choose to change the path of analysis, the computer should be able to shift its presentation cleanly and without delay.

Ultimately, professional users of medical imaging desire easy access to a subset of information pertinent to their investigations. The workstation must respond quickly to changing demands and needs as each user interacts with it. This adaptive machine may actually change over time, "remembering" successful interactions and applying similar approaches to later interactive environments. The machine is a communication link between the professional and subject, as well as between the professional and other experts world-wide. This smart computer will simply do what the user would like to do if he had the time and capacity to access such immense amounts of data. He would gather, analyze, and simplify the information to best apply to his needs, yet always preserving the excess for later reference. The medical imaging workstation of the future will provide this service for the professional, so that he may concentrate his expertise on the application of the information to his patients, clients, or subjects.

A computer workstation which operates on a complex biological model will first require an organized effort among the sciences to establish the necessary data content and structure. Current approaches to bioimaging and information management systems must become better integrated. For instance, bioimaging is presently derived strictly from geometric information; however, human interaction at a workstation will require correlating geometric location to many other space-filling properties such as functionality, dynamics, and chemical makeup. The imaging must become a communication link for the information management system, providing not only display, but an interactive mechanism for the user to direct the flow of information to best serve his needs. Imagine an expert at a cadaver with a student; to communicate a desire for more information the students simply points to a structure, effectively specifying his interest. The expert then focuses his attention, references his own experiences, and then verbally communicates details with the student. This bidirectional communication is a necessity for the imaging workstation. Three-dimensional displays, tactile gloves, stereovision goggles, and many other devices will provide useful methods for effective user interaction with the computer. Audio communication would be convenient, but textual communication is an acceptable alternative.

The future biological imaging workstation needs to be more than a displayer of pretty pictures. The boundaries between the sciences is becoming fuzzier, creating the need to handle overwhelming amounts and diverse forms of information. The current use of standard publications is relatively inefficient in presenting such material in an organized fashion. In this age of information, the academic and professional biomedical communities will become increasingly dependent upon computer systems for diagnosis, treatment planning, reference, education, communication, and transfer of technology.

Issues in the Design of Workstations for Psychology Experimentation

L. R. Bartram, K. S. Booth and W. B. Cowan

Abstract

Three issues must be considered in the design of a computer-based system for psychology experimentation: flexibility and generality, data logging and temporal precision, and stimulus presentation. This paper discusses these issues and presents solutions to them based on the design of a graphics workstation used for running experiments in perception, human factors and cognitive psychology in the Computer Graphics Lab of the University of Waterloo.

1. Introduction

Computer workstations are well-suited to psychology experimentation because of their inherent flexibility, graphics capability and potential for precise timing and control. This paper discusses three issues in designing graphics workstation-based systems to develop, run and analyze cognitive psychology experiments: flexible software architecture to support prototyping and refinement; data logging and temporal precision techniques; and correctly timed stimulus presentation. We draw examples from our experience in designing the psychology workstations now in daily use in the Computer Graphics Lab of the University of Waterloo (CGL).

The CGL psychology workstations uses a dedicated multiprocessor VMEBus architecture running Harmony, a multitasking operating system for real-time control [Forsey85, Gentleman83]. Because Harmony is an embedded system without archival storage or user interface services, the experiment software, specifications, and data files are maintained on a host computer, currently a DEC VAX8600 running BSD 4.3 UNIX. Communication is through Ethernet. Harmony provides lightweight processes (*tasks*) which communicate using efficient synchronous message-passing primitives. A graphics library supports 2D primitives, image and display manipulation using up to three eight-bit frame buffer video boards [Fowler88]. Programs are written as sets of concurrently active communicating tasks.

2. Implementation Overview

Integrated environments for experimentation must support all aspects of the experiment cycle:

1. design
2. specification (structure of events and data that affects them)
3. trial running
4. data collection
5. data analysis

Stages (1), (2) and (5) require a broad software base of tools for specification generation, data randomization and statistical analysis. Steps (3) and (4) depend on a responsive environment where timing can be controlled with millisecond accuracy [Bartram88]. The CGL design solves the dichotomy between richness and efficiency by using the dedicated multiprocessor workstation for experiment running and data logging. Experiment development and analysis is performed in the richer VAX multi-user environment, taking advantage of program development facilities, text editors and sophisticated image processing tools for stimulus preparation [Paeth86].

The software supporting the development and execution of experiments must be flexible enough to provide rapid prototyping of experiments. Flexibility is crusial because an experimenter refines an experimental design by exploring many variants in the pilot stage. A system which is easy to use for those inexperienced in programming encourages full use of that flexibility [Butler88]. Generality is a necessary property to reduce programming overhead: software developed for one experiment is ideally re-usable in other designs. The software component of a CGL experiment is a collection of tasks (*trial daemons*) which embody the experiment activities, administered by a run-time controller that interprets an experiment specification in a specialized description language [Bartram89]. Section 3 describes the advantages of this design.

Data collection is an essential component of any experiment. Any event that occurs, and usually the time that it occurs, can be important data in a given experiment, but a large amount of inappropriate data creates difficulties for analysis routines. An effective data logging scheme can be created on the basis of two principles: every event is time-stamped at its source, but events are only logged when requested by the experimenter. The details and advantages of such a scheme are described in Section 4.

Another essential feature of any experiment is precise stimulus calibration. This problem is well-understood in the chromatic domain, and known to be difficult but tractable in the spatial domain for CRTs [Cowan87, Brainard89]. Temporal calibration, because CRTs provide discrete stimuli in the temporal realm, is more problematic. Unlike data logging, stimuli are best timed by counting the discrete presentations (frame times). The details of stimulus timing are described in Section 5.

3. Generality and Flexibility

Computer systems for psychology experimentation generally fall into one of two camps: libraries of specialized experimentation routines to be invoked from high-level programming languages [Edgell89, Lane87], or "experiment generator" programs, which produce executable code from forms or templates filled out by the experimenter [Costin88, Poltrock88, Schneider88].

The subroutine approach is more powerful and extensible in that it is not restricted to one particular type of experimental design [Edgell89]. However, the experimenter must be familiar with computer programming and operating systems or have good programming support. Experiment generators are easier to use, freeing the user from the details of computer programming. Prototyping and incremental refinement of an experiment involve altering experiment-specific forms rather than re-coding. Such systems have two drawbacks. First, they are limited in the types of experiments supported. For example, all systems surveyed for this paper support only discrete-time experimental paradigms and only one permits the experimenter to specify any concurrent activities [Schneider89]. Second, they tend to be microcomputer-based and correspondingly constrained in computing power and graphics capability. In both types of system, experiment prototyping and testing require re-specification and re-compilation, although in the case of some experiment generators the time involved for recompiling is minimal [Schneider88]. The need to recompile each time another variation of an experimental design is ready reduces the flexibility of rapid prototyping and limits the extent to which different paradigms can be executed on the same compiled software.

The advantages of both approaches are combined by providing a broad base of software tools coordinated by an experiment run-time controller, a virtual experiment machine that interprets a high-level experiment description language [Bartram88]. The specialized language alows the experimenter to focus his or her attention on the experimental concepts by hiding the details of the computer implementation. Such a language is most powerful and effective when it combines three properties: simplicity, compactness and expressiveness [Shneiderman87]. Meaningful command names, simple syntax and a structure reflecting the components of experiments to be designed are all important. If the language is compact, with a small set of syntactic rules that provide no excess functionality, design tools can help the experimenter by detecting erroneous designs interactively. Within these constraints the generality characteristic of a fully expressive language is still possible, mediated by a close correspondence between the semantic concepts of the language and the behaviour of a specialized experimental workstation.

The CGL workstation is based on these concepts. Experiment activities are embodied in trial daemons. Each daemon type is responsible for a well-defined set of experimental activities. For example, general display operations are handled by a `Screen_manager` daemon, text displays by a `Text_writer` daemon, and signaling the end of a trial by a `Termination` daemon. Each daemon accepts instructions telling it what do accompanied by a list of parameters. The commands in the experiment description language reflect the structure of the experiment activities by permitting the experimenter to control the daemons that perform them. The language syntax is compact, encompasses only three command types and is similar to Pascal, a well-known programming language. A textual format and meaningful command and parameter naming conventions make it human-readable. The experimenter lists commands to the run-time controller to tell it what daemons to create, which instructions to dispatch to them and how they interact, and the order in which the activities should occur (sequential or concurrent).

These commands are used in a specification script whose organization into three *command blocks* corresponds to the structure of the experiment. The experimenter begins by listing the daemons to be used in the experiment in the `daemons` block.

```
Screen_manager (dtype=SCREEN_MANAGER)
```

defines a daemon of the SCREEN_MANAGER type named `Screen_manager`. Daemons have unique names so that several daemons of the same type can be used in the same experiment.

A `session` block describes the state to remain constant through the session; it is followed by one or more trial blocks describing the event sequence in that trial. In the `session` and `trial` blocks, the experimenter uses the same two command types to control what activities occur under what circumstances. A *daemon command* tells the run-time controller to dispatch the instruction contained in parentheses to the daemon whose name precedes it. For example, to tell the above display manager to blank the screen named S1, the experimenter would see

```
Screen_manager (cmd=BLANK, scr_name=S1);
```

Each daemon is implemented as a lightweight task in the multitasking paradigm, meaning that it will continue processing independently of other tasks in the system until it requires something from another taks. When it is ready to carry out experiment activities, it requests an instruction from the run-time controller. The controller, in turn, dispatches instructions to the waiting daemons as soon as it encounters in the script. If the controller comes to an instruction destined for a daemon that is not yet waiting, it must wait until the request from that daemon arrives before continuing to dispatch instructions to other daemons. However, if there are several daemons waiting, they can receive new instructions in rapid sequence and be concurrently active.

A method of controlling the simultaneous daemon activity is necessary for synchronizing experiment activities on one or more particular events. This is achieved with a `Wait` command which causes the controller to cease dispatching instructions until the daemon(s) responsible for that event have completed their processing. Thus, by specifying

```
Clock (cmd=DELAY, milliseconds=3000);
Termination (cmd=START, tab_name=Tablet,
               answer=LEFT_OR_RIGHT_BUTTON_PRESS);
Wait for (Clock or Termination);;
```

the experimenter defines two possible conditions for trial termination: either a timeout or a particular tablet button press. The run-time controller will block until the `Clock` or `Termination` daemon returns with a request for another instruction.

This design has several advantages. From the experimenter's viewpoint, rapid prototyping is a matter of changing a script rather than re-compiling the set of programs. Incremental refinement [Schneider88] is easily accomplished by testing a bare bones script and gradually adding commands. Since the script is stored as a simple text file, any standard editor can be used to create and modify it. The experiment machine is an open-ended system: the language offers the flexibility of running different types of experiments on the same compiled set of programs by altering only the scripts. (For example, specification of a multiple-stimulus experiment in our system will use the same commands and daemons as a single stimulus experiment but instruct them differently.) Many of the existing daemons carry out general-purpose experiment tasks (such as handling input and output devices, timing and display) and can be used in dissimilar experimental designs. This re-usability reduces the dependency on full-time programming support and contributes to the generality of the system.

4. Data Logging and Temporal Precision

Experiment events can be associated with peripheral devices, subject input, timing controls, or defined by the experimenter. Our experience has shown us that recording every event that occurs in an experimental is impractical. It produces unmanageable quantities of data and overloads the experiment execution with logging functions, interfering with the temporal calibration of the stimuli. Yet, since any event in the system is a potential candidate for logging, a mechanism must exist to measure it precisely.

Some systems support a fixed data logging scheme in which the types of data to be logged are selected from a predefined list of experiment variables [Costin88, Edgell89]. A more flexible method puts data logging under full control of the experimenter but provides a low-level routine to timestamp an event as soon it occurs. This separates precise measurement of an event from the act of recording it. The timestamp is associated with event data as it is distributed in the system. The decision to record an event is under the experimenter's control and is specified in the experiment description. Logging may thus happen well after an event has occurred but accurate reporting of the event is assured. In the CGL workstation, daemons responsible for particular events also accept commands to log the events. In addition, a special data logger daemon is used to manage general purpose logging utilities such as timer granularity, log file initialization and event naming.

Logging involves access to archival storage (whether file system, disk or host computer). However, the operations for direct access to archival storage are dependent on the speed of the storage devices themselves and cannot be guaranteed to complete within a specific time constraint. This problem is best resolved by using a server task that buffers event data to be logged and uploads the data to the storage system when the tasks carrying out the experiment are not using the CPU. The server accepts a request to log an event from a task in the experiment, queues it and immediately frees the requesting task to continue working. It attaches sequence numbers to the event data and records them in the desired format. Sequence numbers are useful in cleaning up faulty data. Since logging itself is an event, it is possible to have the server timestamp an event when it is logged, so that subsequent inspection can determine whether too much time elapsed between the time the event happened and its recording. In this way, the recording of the events is done "off-line" from the actual experiment events; precision is ensured by timestamping the event when it occurs.

This method permits user-defined data logging while ensuring that the temporal accuracy is maintained. Error handling is supported; the experimenter can decide to log only "proper" responses, or to flag erroneous events and log them as distinct types. We anticipate that this will contribute to improved data logging techniques which increase the resolution of the collected data [Chionh88].

4.1 Temporal Precision Issues

Cognitive psychology experiments require timing accuracy of millisecond granularity [Chionh88, Costin88, Dorfman87, Kieley89]. Temporal precision is mandatory for effective timestamping services. Methods for temporal precision at millisecond granularity include special modification of hardware clocks [Schneider88], software clocks [Dorfman87], custom machine language routines [Schneider88, Lane87] and using the vertical retrace interrupt signal as an interval timer [Costin88, Dorfman87, Lance87]. An architecture based on a time administrator backed by adequate computation makes these approaches unnecessary. As realized in the CGL workstation, a dedicated timer is used for timing operations, ensuring that all system timing references are consistent. The granularity of the timer can be set in the experiment specification to a value between 0.417 ms and 5.208 ms. For efficiency this timer resides on a lightly loaded processor. The timestamp routine simply reads this hardware clock. An administrator task [Gentleman81] handles all details of setting up the hardware clock (based on the granularity specified in the experiment and the processor configuration). One way of controlling the timing of experiment events is provided by a `Clock` daemon which accepts a command to `Delay` for a specified duration. It then blocks on a server which reads the timer and frees the blocked task at the end of the interval. This architecture combines the advantages of other approaches mentioned previously: the efficiency of hardware timers, and the generality of software timing routines.

5. Display Timing

Temporal granularity of stimulus presentation is limited by the frame time of the CRT display (usually 33.33 or 16.67 ms). To avoid observable artifacts on the display, operations that affect the screen (such as setting registers or changing LUT entries) must be performed in synchrony with the frame rate, usually the video refresh time [Fowler88, Chionh88] – approximately 1.3 ms. Timing control in this context is best measured in terms of vertical retrace intervals to synchronize experiment presentation with the display hardware.

Use of a special routine as a Vertical Retrace Server facilitates the synchronization. In the CGL workstation, a task accesses the display by sending a request to the Vertical Retrace Server specifying when the operation should be carried out: immediately, during the next vertical retrace interval, or during some specific interval in the future [Fowler88]. The request is queued up by the server to be carried out at the specified time. The requesting task can continue processing without waiting for the vertical retrace event.

If the hardware generates an interrupt for video refresh the timing is simplified as long the overhead of task context switching for the interrupt service routine does not take up too much of the vertical retrace interval [Chionh88]. This approach is often used for presentation timing in microcomputer systems [Costin88, Poltrock88, Schneider89]. Without a hardware interrupt it is common to busy-wait between intervals [Dorfman87], making execution of other activities impossible. A more efficient approach implemented on the CGL workstation uses a special timer set to generate an interrupt approximately 0.5 ms before the vertical retrace. The interrupt "wakes up" a worker to the Vertical Retrace Server which then busy-waits on a status bit until the retrace begins [Chionh88, Fowler88]. This logical interrupt scheme avoids the overhead of busy-waiting during a complete display frame but reduces efficiency in two ways. First, the worker must still poll the vertical retrace register for 0.5 ms with all interrupts disabled. Having multiple processors, it is feasible to devote a processor to display synchronization and updates. Second, access through a server creates a bottleneck for display operations. Too many requests for screen updates can mean that not all operations will be complete by the end of the interval, again resulting in observable artifacts. Containing the execution of such operations within a specific interval is a yet unsolved problem in real-time system design. Approaches under investigation in CGL include techniques of overlapping and prioritizing requests [Chionh88].

6. Discussions and Conclusions

The initial development of the CGL workstation focused on the three issues discussed in this paper and concentrated on the features of the run-time environment. The next step in increasing the flexibility and power of the system is the addition of specification and debugging aids for the experimenter's use. Although human-readable in format, the experiment description language is inadequate as a user interface and a more comfortable graphical interface is planned. Extensive debugging and verification tools are required, particularly for experiment timing. The latter are well-known problems in real-time systems with no established solutions.

Nonetheless, the CGL workstation design shows that even rudimentary solutions to these problems are effective for useful psychology experimentation. Experiments which have been or are currently carried out in CGL include a mental-rotation experiment using Shepard-Metzler cube structures [Shepard71], an facial recognition task using digitized colour images, and a series of visual search experiments investigating the use of colour. Under development are double-stimulus experiments in binocular vision, and adjustment experiments in tone and colour matching. Common to all these experiments are certain general-purpose daemons, reducing programming requirements. The visual search experiments were designed by a non-programmer using existing daemons and required no programming support at all: the experiment designer merely generated a new specification script.

Data logging formats used in the mental rotation and image recognition experiments are contributing to research into data collection techniques. Excessive reporting of spurious data has led to a refinement in the data logging system for error handling. Finally, both methods of strictly controlled stimulus presentation have proven effective. Specifying a duration of n milliseconds before clearing the display was used in the mental rotation and visual search experiments; defining duration as a number of display frames was preferred in the facial recognition task.

References

[Bartram89] L. R. Bartram. Configurable Multitasking: A Method of Experiment Prototyping in A Multiprocessor Environment. M. Math Thesis, Department of Computer Science, University of Waterloo, 1989

[Bartram88] L. R. Bartram, K. S. Booth, W. B. Cowan, J. D. Morrison and P. P. Tanner. A System for Conducting Experiments Concerning Human Factors in Interactive Graphics. *Proceedings of Graphics Interface '88*, pages 34–42, June 1988

[Booth85] K. S. Booth, W. B. Cowan and D. R. Forsey. Multitasking Support in a Graphics Workstation. *Proceedings of the 1st IEEE International Conference on Computer Workstations*, pages 82–89, 1985

[Brainard89] D. H. Brainard. Calibration of a Computer Controlled Color Monitor. *Color Research and Application 14*, pages 23–24, 1989

[Butler88] D. L. Butler. A critical evaluation of software for experiment development in research and testing. *Behavior Research Methods, Instruments & Computers* 20 (2), pages 218–220, 1988

[Chionh88] E. W. Chionh, K. S. Booth, W. B. Cowan and P. P. Tanner. Multiprocessing, Multitasking, Realtime Graphics Workstations for Psychology Experimentation. *Proceedings of the International Conference on Computer Graphics*, Singapore, Sept. 15–16, 1988

[Costin88] D. Costin. MacLab: A Macintosh system for psychology labs. *Behavior Research Methods, Instruments & Computers* 20 (2), pages 197–200, 1988

[Cowan87] W. B. Cowan. Colorimetric Properties of Video Monitors. *Notes for a short course given at the Annual Meeting of the Optical Society of America*, Rochester, N.Y., 1987

[Dorfman87] D. Dorfman. Software timing of events in cognitive psychology experiments. *Behavior Research Methods, Instruments & Computers* 19 (2), pages 185–190, 1987

[Edgell89] S. E. Edgell and S. A. Hertel. Running laboratory experiments using the RSX operating system and FORTRAN on the PDP-11. *Behavior Research Methods, Instruments & Computers* 21 (2), pages 303–306, 1989

[Forsey85] D. R. Forsey. *Harmony in Transposition: A Toccata for VAX and Motorola 68000.* M. Math Thesis, University of Waterloo, 1985

[Fowler88] P. P. Tanner, B. M. Fowler and K. S. Booth. Experience with Graphics Support for a Multiprocessor Workstation. *Proceedings of Conference on Parallel Processing for Computer Vision and Display*, Leeds 1988

[Gentleman81] W. M. Gentleman. Message Passing Between Sequential Processes: The Reply Primitive and the Administrator Concept: *Software Practice and Experience* 11, pages 436–466, 1981

[Gentleman83] W. M. Gentleman. *Using the Harmony Operating System.* Technical Report NRCC-ERB-966, Division of Electrical Engineering, National Research Council of Canada, December 1983

[Kieley89] J. M. Kieley and T. S. Higgins. Precision timing options for the Apple Macintosh family of computers. *Behavior Research Methods, Instruments & Computers* 21 (2), pages 259–263, 1989

[Lane87] D. M. Lane and B. Ashby. PsychLib: A library of machine language routines for controlling psychology experiments on the Apple Macintosh computer. *Behavior Research Methods, Instruments & Computers* 20 (2), pages 246–248, 1987

[Paeth86] A. W. Paeth. Design and Experience with a Generalized Raster Toolkit. *PROCEEDINGS, Graphics Interface '86,* Vancouver, B.C. May 1986

[Poltrock88] S. E. Poltrock and G. S. Foltz. APT PC and APT II: Experiment development systems for the IBM PC and Apple II. *Behavior Research Methods, Instruments & Computer* 20 (2), pages 201–205, 1988

[Schneider88] W. Schneider. Micro Experimental Laboratory: An integrated system for IBM PC compatibles. *Behavior Research Methods, Instruments & Computer* 20 (2), pages 206–217, 1988

[Schneider89] W. Schneider. Enhancing a standard experimental delivery system (MEL) for advanced psychological experimentation. *Behavior Research Methods, Instruments & Computers* 20 (2), pages 240–244, 1989

[Shepard71] R. N. Shepard and J. Metzler. Mental Rotation of Three-Dimensional Objects. *Science* 171 (3972), pages 710–713, 1971

[Shneiderman87] B. Shneiderman. *Designing the User Interface: Strategies for Effective Human-Computer Interaction*, 435 pages. Addison-Wesley, Inc. 1987

Protocol Toolkits for Distributed Real-Time Experimental Workstations

A. S. Brown

Abstract

An experimental workstation is often an end system in a network of systems generating, processing, and presenting data. Graphic software standards for workstations often extend the virtual display device model but leave little flexibility for application programs using task and process concurrency in all systems, including the workstation. Such applications are true distributed systems; they benefit from use of standard services in communications facilities, and from formal methods of design and implementation for custom protocols, where standard services are inadequate or non-existent.

As an example, a protocol for synchronization of clocks by exchange of timebase messages (based on an existing network timestamp protocol) is generalized to a method for tracking and synchronization of other real-time databases, which can be applied to distributed interactive world-coordinate clipping of a real-time data stream for graphic display. This method of distributed clipping is illustrated in a system for interactive display of real-time weather data.

1. Diversity of Real-Time Experimental Workstations

Graphic workstations are standard laboratory equipment, providing interactive processing and display of data, often in real time and in conjunction with control of experimental apparatus; a laboratory researcher either calls on a software "instrument-maker", or is his own software designer and programmer - and graphics facilities are an essential part of the toolkit. As examples of the wide range of workstation applications in experimental systems, consider the papers presented at this conference. In many cases, such as the IBIT distributed process control system [Koegel89] and the University of Waterloo system [Bartram89] for experiments in visual perception and cognitive psychology, several microprocessors are used to separate non-critical functions from measurement and control, to achieve precision and responsiveness in time. These are true event-driven real-time distributed systems, although communications functions are not highlighted in their discriptions.

In many cases, such distributed systems can take advantage of existing communication protocol tools, from link protocols (providing reliable point-to-point transfer of fixed-length blocks) to transport protocols (providing reliable network transfer of messages of any size), and tools built on them, such as distributed graphics systems (e.g. X-Windows) and network file systems. In other cases, unique requirements and unusual communication hardware may call for custom-designed protocol systems. In either case the software instrument-maker can add standard protocols and methods of protocol design and implementation to the toolkit. This paper provides a tutorial introduction to some of these facilities and methods.

1.1 Software Models: Concurrency

Event-driven programming is common in user-interactive PC software, which is often called "real-time" in the sense of responding smoothly to user keyboard/ mouse actions, to provide a WYSIWYG screen interface. Events originate in this small fixed set of interrupting devices, fully supported by the operating system; and although rapid processing is needed, the "real-time" constraint is a rather loose one, of user perception. A single-threaded loop programming style, perhaps using queueing of event tokens, is a natural way of writing this type of software. The Macintosh computer's operating system, for example [Mac87] enforces this model for all programs that use the standard window and icon graphics interface, and imposes a hierarchy of event priorities within this model, providing concurrency only at the level of interrupt service; MS-DOS in PCs provides even less in the way of event management services.

While the single-threaded loop was common practice in older real-time systems, and is still typical as a "cyclic executive" or schedular in embedded processors [Shaw86] most real-time systems now assume concurrency in multitask or multiprocess programming (where "tasks" share an address space but "process" do not) as well as between interrupt and base level execution. Exchange of information between these concurrent software units has been thoroughly discussed and is well understood. The literature's classics are [Dijkstra68], [Hoare78], and [Hansen78]. A survey of these and other approaches is found in [Carson82].

Consistent and reliable design of software systems built from such units, however, remains something of an art. As an example, an attempt to formalize the problem of "deadlock", or mutual exclusion, among concurrent processes, is found in [Gilbert72], which requires description of a distributed system as a finite state machine composed of all the states of all its concurrent processes; this massive effort is rewarded with a computational method of proof of correctness (i.e. absence of deadlock) which is rarely useful in practice.

1.2 Message Oriented Distributed Systems

W. M. Gentleman points out that we naturally tend to think of independent, concurrent software units as anthropomorphic, and can use this inclination to structure real-time concurrent software effectively:

... essential in the initial stages of problem decomposition and structuring for concurrency and parallelism. There is ample opportunity for establishing mathematical properties and proofs after the structure is chosen. All branches of science and engineering rely heavily on analogies and visualization to prompt insight and intuition, and the closest experience most of us have with collections of asynchronous processors is with organizations of people. Moreover, organizations of people function effectively (though perhaps not ideally) with a much higher degree of parallelism than programs anyone has yet been able to build. ... [Gentleman81]

This approach to software is indeed widespread, as indicated by the common use of the terms *customer, server,* and (perhaps stretching the anthromorphic analogy) *daemon,* to describe common roles in concurrent software design:

- *Customers* enqueue requests for service requiring some resource;
- One or more *servers* provide that service and resource to customers, one at a time;
- *Daemons* are created to magically and secretly carry out instructions concurrently with other customers', servers', and daemons' actions, and destroyed when their work is done.

These relationships are meaningful and perform useful work, through exchange of messages, often organized in anthropomorphic fashion, if not natural language; the queueing systems that result, moreover, are well understood in theory and practice. As might be expected, however, the complexity of relationships that can arise can result in failures of design. A *sorcerer's apprentice,* for example, may continuously create daemons that refuse to die!

1.3 Language or System Features

To improve consistency of design and to make analysis by *proof of correctness* feasible, programming languages from Fortran to ADA have been augumented or designed with features for management of concurrency; and operating systems from Multics to Mach have provided models for concurrency. The most widespread of the system features may be the UNIX *fork* and *pipe,* but these are among the least useful for real-time systems. The most widespread concurrent language at this time is probably ADA; the most widespread real-time operating system is probably Intel iRMX.

While integration and concealment of task/process creation, control, deletion and message facilities in language and system environments provides the most disciplined approach to software design, it rarely solves problems of software design for heterogeneous systems made up of a mix of very different processors – the typical situation in a distributed experimental system. The software instrument-maker usually chooses languages and systems suited to specific needs, designing message oriented solutions composed of separate programs in different languages (such as C, for ease of UNIX development, and Forth, for ease of interactive real-time development) to run in different processors linked with communications facilities, rather than a single program using integrated language or system facilities.

2. Distributed Systems Models

Any convention for exchange of messages that relate events in distinct concurrent systems forms a protocol. The software instrument-maker is thus either a user or a designer/implementer of protocols that exchange messages over communications facilities. Wherever standard protocols can fill the needs, of course, they are far more economical than custom protocols.

2.1 ISO Open Systems Interconnection (ISO/OSI) Standards

Computer networks blossomed from the DoD ARPANET project and the TCP/IP protocol suite, which was developed in part by trial and error in a nationwide research program in the 60s and early 70s. (See [Comer88] or [Stallings88] for an introduction.) For a variety of reasons, including its somewhat informal origins, the TCP/IP suite was resisted as an international industry standard for network systems. Since 1975, a powerful movement toward standardization has developed at the interface of public packet switch networks, and local area networks (LANs). This process began in national and international technical organizations, especially the International Standards Organization (ISO), but gained significant momentum when endorsed by the US Federal government as a means of controlling computer procurement costs.

In this standardization process, *network* is used to describe any of a variety of methods for providing program-to-program and human-to-program data interfaces among a set of general-purpose computers with roughly equivalent capabilities, differing only quantitatively. A local area network of workstations and minicomputers is a good example of such a network. This *peer-oriented* network is in contrast to a *hierarchical* network of special-purpose or limited-function machines surrounding a general-purpose computer.

As a result of the ISO "Open Systems Interconnection" program a model of organization for network communications systems was accepted by nearly all vendors and users in the computer industry. This model identifies seven separate functions or *layers* to be provided wherever complete communications services are required between two computers. Each layer uses the services of the layer immediately below, to add a particular facility; this *nested* organization makes upper layers independent of details of lower layers. Most products use software (often in ROM firmware) to implement layers above the first two.

The seven layers are in three general groups, relating to specific physical communications equipment, an abstract communications service, and general-purpose *end-to-end* software facilities. (In particular interfaces, an unneeded layer may be absent. The only layer typically present in every interface is the Transport Layer, which provides the universal communications service required for Open Systems Interconnection.)

Physical and logical equipment-related services:

- *Layer 1, Physical:* electrical and mechanical standards.
- *Layer 2, Data link:* standards for behavior of logic and software to move fixed-size data units without error across a single connection.
- *Layer 3, Network:* standards for behavior of software to move fixed-size data units across a network of many connections without error.

Abstract communications service:

- *Layer 4, Transport:* standards for behavior of software to move arbitrary-size data units without error across any network or combination of networks – an abstract communications service.

End-to-end general purpose facilities:

- *Layer 5, Session:* standards for identifying and connecting to partners in a communication exchange, for controlling the order of "speaking" during the exchange, and for synchronizing and marking the exchange of information, esp. in order to recover from network failure during exchange.
- *Layer 6, Presentation:* standards for encoding information, including compression and encryption.
- *Layer 7, Application:* standards for general purpose facilities such as file transfer, virtual terminal service, and electronic mail. Also includes facilities for program-to-program communications and sharing of file systems and other processor resources.

This model and its terminology was developed primarily for application to general-purpose timesharing and workstation systems, and is rarely used in its entirety for laboratory data acquisition and other real-time systems. However, the description of *layered* hardware and software functions is still appropriate.

2.2 Layered Communications Protocols

Standard facilities are usually available at only a few of the seven levels in the ISO Open Systems Interconnection model (or the equivalent DoD TCP/IP protocol suite); most protocol *stack* implementations do not offer easy programmer access to lower or internal layers. The most common levels of access are:

- *Link layer* (SDLC, HDLC, ADCCP, etc) – point-to-point link connect; link data frame send/receive request, notification and confirmation; link disconnect
- *Transport layer* (TCP, ISO TP4; UNIX BSD *socket,* SV *stream*) – network address connect (multiplexed within processor); message or "transport service data unit (TSDU)" send/receive request, notification and confirmation; network disconnect
- *Session layer* (Netbios, NFS-YP) – network name connect; message authorization/validation; message request/grant; message send/receive request/notification/confirmation with optional checkpointing; network disconnect
- *Application layer* (ISO CASE, FTAM, VTAM, MHS; DoD TFTP, TELNET, SMTP) – application name connect; access authorization/validation; session/presentation services; file transfer services; virtual terminal services; E-mail services; application disconnect.

Some environment also provide access to the network layer, e.g. X.25 packet level protocol, or DoD RFC768 UDP: User Data Protocol. Use of DoD application services is limited to ability of the user program to reproduce protocols; source code is often available with the UNIX OS.

2.3 Common Application Services

Appropriate services for real-time experimental applications range from non-existent to abundant, when available, over these layered protocol facilities:

- *Physical layer* net boot service
- *Link-layer* shared-buffer *snapshot* exchange service
- *Master-slave* operation of remote processors (e.g. Intel BITBUS)
- *Network File System* – file open/close, read/write server in each processor's OS; issues file system access requests (e.g. Sun NFS; AT&T RFS)
- *Remote Procedure Call* – used over TCP/IP to provide Network File System, and other services (e.g. Sun RPC)
- *Remote breakpoint/debugging services* (various proprietary real-time products)

2.4 RPCs in X-Windows Limit Concurrency

The most general and powerful of these facilities is certainly Remote Procedure Call (RPC), a method for providing a server on a dedicated processor with a protocol for transporting call-by-value parameter lists for execution of procedures on behalf of a network client. In many cases, even RPC facilities will not be adequate to implement a distributed real-time system, often because of the difficulty of implementing full RPC facilities in a foreign processor, and often because RPC facilities themselves are inappropriate. In X-Windows [O'Reilly88] for example, an RPC protocol is used to provide a host processor with a graphics subroutine library (comparable to GKS) which accesses a display viewport (*window*) on a bitmapped workstation display, through remote procedure calls to a server process (e.g. *uwm*) running on the workstation, which is dedicated to management of this viewport and other *window* sessions on the display.

This set of services provides an extension of the virtual terminal model of graphics system standards such as GKS, to terminals accesses through a network; but the penalty is the restriction of local processing at the workstation to the plotting functions involved in emulating the virtual terminal. Moreover, the RPC-calling host is required to block the calling process until the RPC is completed or acknowledged. This severely limits the concurrent processing that is possible in the workstation; concurrency is achieved only in the host, where several such RPC-calling processes may be executing during each others' blocked periods.

If the RPC-calling process is in the workstation itself, the virtual terminal abstraction is maintained, at the great performance expense of communications system overhead in each graphic operation transaction, usually unacceptable in real-time systems. In addition, the volume of data transferred between host and workstation is sometimes significantly increased by the transfer of graphic-display information rather than real-world event information.

2.5 Need for Custom Protocols

For these reasons and many others, the software instrument-maker often finds himself a software toolsmith (or even *flint-striker*) - creating protocols for unique problems. When this happens, the tools developed in the course of the evolution of network systems are extremely useful: principles of layering, message structure design and implementation, methods of formal specification of software behavior, and tools for implementation of control of software behavior.

2.5.1 Principles of Layering

There are often situations in which several different types of data are passing through one link - for example, a real-world data stream, combined with instrument status, communications link status, instrument/actuator control, and operator/system control commands. The lesson of OSI layered structure, in this context, is the great value of separation of qualitatively different information, by multiplexed addressing to distinct, concurrent destinations. In a real-time system environment where concurrent tasks/processes are available as destinations for messages, an application-layer addressing scheme can be set up for distribution of messages by type code, even if a full transport-layer protocol is not available to provide multiple "sockets" or service access points within a processor. This principle can also distribute messages within a process; e.g. as requests handled by daemon tasks within a server process.

The principle of layering leads naturally to the *envelope* structure of messages: information addressed to a specific destination is prefixed with address and control information, and optionally combined with other addressed information into a higher-level package, which is in turn prefixed with address/control information, etc. The information fields of a top-level message can be passed directly among tasks within a single process address space, and in some cases can be passed between processes without copying.

The layering principle does, however, tend to introduce an additional *postman* task or process providing demultiplexing of message traffic. To minimize scheduling overhead of this demultiplexing, tasks or other *lightweight processes*, or a common address space, are highly desirable. When concurrent processes are not available, a system of queues and an *event manager* procedure within a single threaded loop (in the manner of Macintosh DOS) can often suffice.

2.5.2 Message Design

Within a message data field, *presentation-layer* tools for data encoding and compression can often save valuable communications bandwidth:

- *Huffman encoding* – ordering of codes by frequency of occurrence, with assignment of increasing numbers of bits to less-frequent
- *Run-length encoding* – replacement of a block of identical codes with a run-length code block consisting of a flag, a single value and a repeat count
- *Syntactic encoding* – replacement of sequences of codes with codes representing these units
- *Semantic encoding* – replacement of whole messages with codes representing their value in the context of preceding and following messages – e.g. on a reliable virtual circuit, transmission of data change information rather than instantaneous value, for slowly changing measurements. (NB: requires protocol for initialization.)

The greatest *density of information* in each message, and thus the highest level of compression and greatest communication efficiency, is always found in semantic encoding, where a stream of event information causes synchronization between two *meaningful* representations or models of some observed reality. This, of course, effectively creates a distributed database consisting of the redundant model representations; it also requires the greatest *intelligence* in the application protocol software, in the sense of embedded information about the real-world phenomena involved at both sending and receiving stations, and is least susceptible to standard tools. There is, however, a large set of common problems in managing the behavior of such a distributed system, which is indeed susceptible to a toolkit approach.

2.5.3 Formal Specification of Behavior

Two formal structures have been found useful for specifying protocol behavior in concurrent systems: Petri nets and finite state machines.

- Petri nets are most useful for describing synchronization properties of message exchange. Each concurrent system is depicted as the path of a token along a game-board consisting of *nodes*, or places where the token stops in discrete time, and *transitions*, or places where the token continues after some condition is satisfied, usually arrival of one or more additional tokens at the transition. The model was devised to represent hypothetical neural networks, and the transition is said to *fire* like a neuron when the condition is satisfied. The exchange of a message is represented by transfer of a token to a node or transition. The network topology and the initial token conditions fully specify the behavior of the system in discrete time, although formal analysis is difficult. (Some conditions, such as internal events or timeouts, are difficult to represent within the model.)
- Finite state machines (FSMs) are useful for describing action sequence properties of systems during message exchange. (The usage and sense of state machine notation is closer to that of the digital systems engineer than the linguistics theorist; a *regular grammar* is rarely a useful concept.) A FSM consists of a network of states, joined by transition arcs, which are followed based on the value of mutually exclusive conditions associated with each arc leaving a node. As in Petri nets, a *token* is moved through the network from state to state to represent changes in activity of the system being described. A state may have a new FSM associated with it, describing a *mode* of operation; *internal* events in this *nested* FSM may be the cause of a transition out of the current state.

Both methods are susceptible to computed formal analysis for reachability and deadlock, but for most practical cases these are obvious by inspection if the diagrams are carefully drawn. Petri nets and FSMs are complementary in the sense of portraying different aspects of the same behavior: Petri nets are excellent for describing synchronization behavior, but less clear in describing sequencing; FSMs are best for describing sequencing but less clear in describing synchronization. A protocol design is best worked out by beginning with a rough design in Petri net notation, followed by FSM elaboration.

The FSM can only accomodate moderate levels of detail before the diagram becomes unreadable. To record full detail (for example of error and exception actions in each state) the FSM can be restructured into a *transition table* noting action and new state for each event in each state. When adequate details is reached, a time sequence diagram (similar to digital logic waveform diagrams) of message exchanges can be constructed to clarify fine points. These four design methods should be combined with a pseudocode layout of implementation, at least for individual actions.

2.5.4 Tools for Implementation of Control

One enormous advantage of FSM formulation of a protocol is the existence of software tools for implementation of FSM parsers, esp. *lex* [Lesk75] in the UNIX Programmers' Workbench toolset. This is now available in several non-UNIX implementations, including a few public-domain distributions (one simplified version useful for small state machines is distributed widely with the standard Columbia University KERMIT protocol for async file transfer [DaCruz87]).

Lex-like parsers rely on the existence of an *input()* function to obtain an event symbol as input to transition processing; this makes waiting for an event, or processing of a substate FSM during a state, somewhat awkward. It is often useful to provide a daemon to execute a substate FSM or block for an external event while input() blocks waiting for a symbol returned on termination. This again indicates a preference for tasks or lightweigth processes.

There are significant differences among lex implementations, and quirks common to all. [Volkman89] Lex parsers generate C-language "switch" statement groups, which require an *int* type (integer) as the symbol code – and *int* types vary in precision in different processors; some simplified lex implementations assume a *char* type. These considerations affect the data type of the token returned by input() - but in general, it cannot be a pointer to a memory area, and queues must be implemented separately.

The output of lex is a C-source module, which may be modified by hand if necessary; the intent, however, is to provide a higher-level language for FSM implementation *compiled* by lex into C. Standard lex generates a state machine function named *yylex()* and does not permit use of more than one lex-processed module in the same C program; recent versions have command line switches modifying the name of *yylex()* and other symbols, so that nested FSMs are more easily used.

3. Distributed real-time Experimental Workstations

These tools are generally useful for design and implementation of custom protocols for distributed real-time systems. Many of these systems can be divided into a sensor-control-actuator data acquisition network, a supervisory controller system, and an experimenter's workstation, linked with communications subsystems.

3.1 Example: IBIT

The IBIT system [Koegel89] illustrates an object-oriented user interface toolkit approach to a sensor-control-actuator system. IBIT provides tools for *end-to-end* representation and manipulation of the objects of this type of system, in which objects are composed of:

1. Sensor data provided by messages from asynchronous data acquisition processes
2. Control outputs to the supervisory controller processes
3. Operator workstation outputs and inputs (display, mouse, keyboard etc.) and
4. Event-driven FSM process and associated data, defining and implementing behavior of a composite representation of a subsystem, from the operator's viewpoint.

Events are interactively defined by the operator, to create an object; objects can be composed of other objects, and associated with an icon in the operator's display: icons can be replicated, connected or overlapped on the display, causing corresponding manipulations of objects.

The description of IBIT in [Koegel89] does not highlight the role of communications interfaces between the segments of this system, because most needs were filled by standard protocols, including X-Windows display facilities.

3.2 Common Problems

The IBIT example, however, clearly shows some of the common protocol problems found in distributed real-time experimental systems that are open to possible modelling in a protocol tool kit. These include:

- sensor data point value message *stream*,
- alarm messages regarding sensor data point monitor with deadband,
- Control messages to sensor/actuator system, including deadband threshold change,
- Timestamping of messages and real-time clock synchronization,
- Hierarchical distribution of real-time database,
- Peer update of distributed data base,
- Event messages associated with FSM transitions.

3.3 Tools

Most of the protocol tools provided by modern network systems, including both TCP/IP and ISO protocol stacks, are appropriate for efficient operation of *peer-oriented* networks in which, at each layer, message exchange may be initiated by any functional entity in the network system. This conceptual model of network systems simplifies protocol design and implementation in many ways, because it presumes (and requires) a single common protocol system present in all computer systems attached to the network.

Peer network systems are not always appropriate or efficient for distributed real-time systems, however, because the full protocol software system in the standard networks is usually large, and designed for efficient service among roughly comparable mainframe and minicomputers and workstations, for the three basic human- user services: file transfer, virtual terminal access, and electronic mail. These network uses are characterized by very bursty traffic with roughly Poisson statistics, and a very wide range in traffic volume; the office network systems designed for them must provide graceful degradation under heavy load, but also have a user population that *degrades gracefully* with declining network performance, because there are few external real-time constraints, if any.

Real-time network systems are usually composed of two or more very different types of processor, and often must support sustained high traffic volume rather than bursty traffic, and can tolerate only limited deterioration in communications delay. Subset peer protocol stacks that don't have these properties, specifically designed for real-time systems, are becoming standardized but are not widespread, e.g. Express Transfer Protocol (XTP). [Chesson88] [Cohn88] Token bus networks provide improvement in these characteristics while supporting the full protocol stack (although at considerable cost) as does an unloaded dedicated office network such as Ethernet. Several proprietary products use lightly-loaded Ethernets to link TCP/IP-based application protocols for real-time systems. For industrial networks, the GM Manufacturing Automation Protocols set based on the ISO network protocols provides the Manufacturing Message Service (MMS) through an application-layer protocol providing network connection to a cell controller, for file or program upload/download, operation start/stop, data point read/write access, and device status. [Kaminski86] These standard and proprietary services may be valuable in some real-time experimental systems.

More often, however, the small dedicated processors typical of real-time systems cannot support these protocols, and economical tools for real-time systems will be found in older *hierarchical* terminal protocol systems, where a *master* station *polls* or grants access to *slave* stations on a shared communications medium; and in methods of multiprocessor system integration, where shared-memory rather than message-exchange interface predominate. It's in these communications environments that protocol software tools are not widely available, and must be custom-built.

Even where full sets of software tools are available, their functions typically satisfy only a few of the needs of real-time systems such as those identified above. Only the last of those, peer update of a distributed database, is solved with a standard protocol, the Committment, Concurrency, and Recovery (CCR) functions in the ISO Application Common Service Element protocols (ACSE). There is, for example, no standard protocol for something as basic as automatically synchronizing local system clocks throughout a network. This problem will be taken as the starting point of the examples below.

3.3.1 Reliable Versus Unreliable Protocols

Perhaps the most basic tools of all network communications systems are the error-correcting protocols for reliable transmission of data over an unreliable link between two systems. In all such *link-layer* protocols a block of data is identified as a transmission unit usually called a *frame*, and a mathematical error-detection algorithm is defined over the bits within the frame that produces the contents of a *frame check sequence (FCS)* field in the frame, usually at the end of the frame. Several families of such algorithms exist, but only a few are chosen for standardization, primarily on the basis of effectiveness in error detection, and ease of efficient implementation. The simplest is a checksum of data contained within the frame (excluding checksum itself). The receiving station then computes the same algorithm over the data, and causes an error condition if the result does not match the received FCS.

This detection of an error is logically quite distinct from actions taken to correct it. In most link protocols, frames received without corruption are acknowledged (ACK'd) either explicitly or implicitly, and corrupted frames are negatively acknowledged (NAK'd), again either explicitly or implicitly. This notification of the error is again somewhat independent of the strategy for correction taken at the transmitter; the missing frame may be retransmitted alone (selective retransmission), or may be retransmitted as the beginning of a back-up (goback-N) process; various combinations of strategies for different error scenarios are also possible.

Choices in each of these protocol design issues define the link protocol. Regardless of these choices, however, the recovery process involves one to possibly many frame times, to recover a single missed frame. This is appropriate in most network system scenarios, where the service requirement is perfect transmission with reasonable performance criteria under loose real-time tolerances.

In many real-time experimental systems, however, reliable link service is neither necessary or appropriate. A data stream of periodic samples, for example, produces a sequence of values, each of which invalidates its predecessor. Perfect transmission of such a data stream is needed under some circumstances (e.g. noiseless recording) but in many situations a missing point or two will not seriously affect downstream processing. The most serious error scenario is failure of the stream to produce sample messages at all – not their occasional corruption. There is thus great value to detecting arrival of every message, but less importance in the validity of its contents. Under these circumstances, while a FCS is still necessary, a simple discard strategy is sufficient on detection of corruption of data, with counter criteria for detection of a persistent link failure condition.

The bulk of a real-time data stream may consist of such uncorrected data messages. Some operations may still require reliable transmission, however – e.g. upload and download of code and data. It is often practical to add reliable service only when necessary, by partitioning off messages for that service to an auxiliary protocol that can provide it, by layering. By this strategy, for instance, a low-speed unreliable async link can carry an uncorrected realtime data stream, command messages corrected with an ACK-NAK protocol, and file transfer messages corrected with a go-back-N protocol, multiplexed together on a single link for different sources and destinations by a custom protocol layering the correcting protocols over an uncorrected frame protocol. In general this can only be done with custom-designed protocol tools.

3.4 Sample Protocols for Synchronization

These examples address simple problems of synchronization between processes in separate computers coupled with unreliable communications links. Synchronization is first addressed as a problem of correcting drift in local time against a timestamp received from a system with higher authority. This method is then generalized as a method of distribution of a database, in a discussion of distributed data stream management, as a form of *clipping.* This approach to generalized clipping is then applied to interactive control at the workstation of the stream of data through a real-time system, from sensor to display, as a generalization of display-window clipping of graphic data. A protocol that can provide real-time control of this end-to-end clipping processes is suggested, with an example of its use in an application to real-time display of lightning weather data using PCs connected to a system providing distribution of weather information over async low-speed dialup lines.

3.4.1 Timebase Synchronization

In a real-time distributed system, a universal timebase is presumed to exist, as provided by a reference standard, and synchronization with it is in principle possible, to any given degree of physical precision. For highest precision, distribution of the reference signal or data stream directly to each end system may be the best solution, but the most economical solution is usually to find a way to distribute timebase information in messages through communications links shared with other data streams. This requires, first, a method of estimation of link delays between systems; second, an algorithm for correction of drift; and third, a method of identifying the authority of a received timebase message.

3.4.1.1 Link Delay Estimation

An established method of link delay estimation exists in the TCP/IP protocol suite. In the Internet Protocol (IP) for routing packets through unreliable networks and subnetworks, the expected delay due to transmission overhead becomes an important parameter in routing and error recovery, because each IP datagram is granted a *lifetime* to reach its next destination, and Internet gateway processors must begin routing recovery algorithms if too many IP datagrams die before reaching their destination. The Internet Control Message Protocol [RFC777, revised in RFC792] for managing routing and recovery processing, provides a service for monitoring link delay by exchange of timestamped messages.

```
0                   1                   2                   3
0 1 2 3 4 5 6 7 8 9 0 1 2 3 4 5 6 7 8 9 0 1 2 3 4 5 6 7 8 9 0 1
+-+-+-+-+-+-+-+-+-+-+-+-+-+-+-+-+-+-+-+-+-+-+-+-+-+-+-+-+-+-+-+-+
|     Type      |   Code = 0    |          Checksum             |
+-+-+-+-+-+-+-+-+-+-+-+-+-+-+-+-+-+-+-+-+-+-+-+-+-+-+-+-+-+-+-+-+
|           Identifier          |        Sequence Number        |
+-+-+-+-+-+-+-+-+-+-+-+-+-+-+-+-+-+-+-+-+-+-+-+-+-+-+-+-+-+-+-+-+
|     Originate Timestamp                                       |
+-+-+-+-+-+-+-+-+-+-+-+-+-+-+-+-+-+-+-+-+-+-+-+-+-+-+-+-+-+-+-+-+
|     Receive Timestamp                                         |
+-+-+-+-+-+-+-+-+-+-+-+-+-+-+-+-+-+-+-+-+-+-+-+-+-+-+-+-+-+-+-+-+
|     Transmit Timestamp                                        |
+-+-+-+-+-+-+-+-+-+-+-+-+-+-+-+-+-+-+-+-+-+-+-+-+-+-+-+-+-+-+-+-+
```

Fig. 1. RFC 792 ICMP - Timestamp or Timestamp Reply Message

Addresses: The address of the source in a timestamp message will be the destination of the timestamp reply message. To form a timestamp reply message, the source and destination addresses are simply reversed, the type code changed to 14, and the checksum recomputed.

Type: 13 for timestamp message; 14 for timestamp reply message. - Code: 0

Checksum: The checksum is the 16-bit one's complement of the one's complement sum of the ICMP message starting with the ICMP Type. For computing the checksum, the checksum field should be zero. This checksum may be replaced in the future.

Identifier: If code = 0, an identifier to aid in matching timestamp and replies; may be zero.

Sequence Number. If code = 0, a sequence number to aid in matching timestamp and replies; may be zero.

Description: "The data received (a timestamp) in the message is returned in the reply together with an additional timestamp. The timestamp is 32 bits of milliseconds since midnight UT. One use of these timestamps is described by Mills [RFC778].

"The Originate Timestamp is the time the sender last touched the message before sending it, the Receive Timestamp is the time the echoer first touched it on receipt, and the Transmit Timestamp is the time the echoer last touched the message on sending it.

"If the time is not available in miliseconds or cannot be provided with respect to midnight UT then any time can be inserted in a timestamp provided the high order bit of the timestamp is also set to indicate this non-standard value.

"The indentifier and sequence number may be used by the echo sender to aid in matching the replies with the requests. For example, the identifier might be used like a port in TCP or UDP to identify a session, and the sequence number might be incremented on each request sent. The destination returns these same values in the reply."

3.4.1.2 Drift Correction Algorithm

The reference to RFC778 identifies a method for using these timestamp messages to distribute a timebase from a single source at COMSAT throughout ARPANET. The same message is used, and an algorithm is provided to compute an estimated correction for link and processing delay:

"Delay measurements can be made with any DCNET host by simply sending an ICS datagram in the above format to it and processing the reply. Let t1, t2 and t3 represent the three timestamp fields of the reply in order and t4 the time of arrival at the original sender. Then the delays, exclusive of internal processing within the DCNET host, are simply (t2-t1) to the DCNET host, (t4-t3) for the return and (t2-t1) + (t4-t3) for the roundtrip. Note that, in the case of the roundtrip, the clock offsets between the sending host and DCNET host cancel."

This provides an estimate, within a few milliseconds, of link delay regardless of drift. The COMSAT system provides a single system with clock synchronized to the US Naval Observatory time standard WWV, and thus any ARPANET system exchanging an ICMP time stamp message with this system obtains an estimate of the true time of receipt of reply within milliseconds, which can be used to adjust the system clock. (No specific algorithm or requirement is provided for doing so. In particular, no delta-t threshold is recommended for a drift requiring correction.)

3.4.1.3 Authority

In many networks, it's not possible to provide a single reference time standard system which can be reached directly from all systems needing timebase synchronization. Timebase information must be distributed through two or more levels of subnetwork system, or among systems that can't reach the reference system for other reasons. In this case, addition of an authority level to the ICMP timestamp message clarifies which of two systems exchanging the message is closest to the reference system, and suggests a peer protocol for clock synchronization. Like the ICMP itself, this protocol can use an unreliable (uncorrected) link.

3.4.1.4 A Tracking Algorithm Protocol for Timebase Synchronization

The *code*, *identifier*, and *sequence* fields of the ICMP timestamp message invite use for adaptations of the timestamp exchange protocol. Consider a protocol using the timestamp message to perform slave tracking of a master clock. Suppose the *identifier* field indicates decreasing authority with increasing numerical level, starting at zero; then the algorithm for drift correction requires that a system receiving a timestamp message adjust its system clock only if the authority level of the timestamp is numerically lower than the current clock authority, and set the new clock authority to the received timestamp's authority plus one. For confidence in correction, the timestamp exchange will be repeated and drift measurement averaged; when average drift exceeds a reasonable threshold adjustment will take place. During this averaging period, the adjusting system's clock remains at its previous value and authority level, and incoming timestamp requests will be rejected with an indication that correction is in progress.

A system correcting its clock for drift according to this algorithm will recheck its clock by exchanging a timebase message again, with the highest-authority system known, after a period equal to the last drift correction (to a reasonable limit of precision), and increase this correction interval by a factor, perhaps doubling, on finding correction unnecessary. Initially, all clocks will be set to the best available local time, authority set to the highest numeric value, and timestamp inquiry of known systems will begin at a default rate. When the network's time initialization stabilizes, each system's clock will track the master system's reference clock to the precision of the minimum correction interval, times the network distance from the reference system, as indicated by the authority level of the local clock. Over a low-speed-serial link, this should permit synchronization within one second.

Parameters for this protocol include:

(bu average drift measurement sample size;
(bu drift correction threshold;
(bu minimum drift correction interval;
(bu valid clock correction interval factor;
(bu default correction interval;
(bu master clock reference system identification.

The protocol can be implemented using the ICMP Timestamp Message structure as identified above, provided only with interpretations of Code and Identifier fields. Code should take a value identifying this clock sync tracking algorithm; Identifier is used as Authority; Sequence may be used to indicate request count during averaging. Nonzero Sequence can also be used to reject an incoming clock sync request from another source, during correction.

```
 0                   1                   2                   3
 0 1 2 3 4 5 6 7 8 9 0 1 2 3 4 5 6 7 8 9 0 1 2 3 4 5 6 7 8 9 0 1
+-+-+-+-+-+-+-+-+-+-+-+-+-+-+-+-+-+-+-+-+-+-+-+-+-+-+-+-+-+-+-+-+
|     Type      |   Code = 1    |          Checksum             |
+-+-+-+-+-+-+-+-+-+-+-+-+-+-+-+-+-+-+-+-+-+-+-+-+-+-+-+-+-+-+-+-+
|           Authority           |       Averaging Count         |
+-+-+-+-+-+-+-+-+-+-+-+-+-+-+-+-+-+-+-+-+-+-+-+-+-+-+-+-+-+-+-+-+
|     Originate Timestamp                                       |
+-+-+-+-+-+-+-+-+-+-+-+-+-+-+-+-+-+-+-+-+-+-+-+-+-+-+-+-+-+-+-+-+
|     Receive Timestamp                                         |
+-+-+-+-+-+-+-+-+-+-+-+-+-+-+-+-+-+-+-+-+-+-+-+-+-+-+-+-+-+-+-+-+
|     Transmit Timestamp                                        |
+-+-+-+-+-+-+-+-+-+-+-+-+-+-+-+-+-+-+-+-+-+-+-+-+-+-+-+-+-+-+-+-+
```

Fig. 2. Clock synchronization message and reply

3.4.2 Database Synchronization: CCR

In the ISO model, computer programs exchange information as "application entities (AEs)" which interact through requests to the ISO Application Layer functions in their computers. These functions are known as "Application Service Elements (ASEs)", and protocols have been defined for a set of "Common Application Service Elements (CASE)". The main service provided by the application layer is unambiguous identification of the purpose of the communications exchange; control of this identification is provided by an "Association Control Service Element (ACSE)", one of the ASEs provided by the Application Layer. After the connection is established using ACSE services, the two AEs can exchange information using presentation layer services (which are requested from and granted by the ACSE). Other services are also provided at the application layer by specialized protocols, including those for the essential human-user services, File Transfer and Manipulation (FTAM), Virtual Terminal Access Method (VTAM), and electronic mail Message Handling Services (MHS).

The CASE protocols include standard services for reliable program-to-program transaction update of a distributed database (essential to all electronic funds transfer systems) through the Committment, Concurrency, and Recovery (CCR) protocols (ISO 8649/3 and ISO 8650/3). CCR provides a program requesting update of a database residing in several locations, each under control of its administrator program, with a handshake procedure for confirming a two-phase *atomic action* that changes all data held to its final state. The first phase of the atomic action permits any participating database writer to reject the operation; if none does, the second phase confirms it and all participants agree on a *committment* to the results. The CCR protocols require the full seven-layer ISO protocol stack for proper operation, although the basic ideas can be extracted for a simpler custom protocol.

3.4.3 Database Synchronization: Tracking Protocol

While CCR concepts can be useful in real-time systems, they require expensive reliable network services. Moreover, the database concurrency problems here are more often like the drift and synchronization problems discussed above: authority relationships are hierarchical and pairwise rather than distributed manyfold; changes occur asynchronously, in small groups of parameter values; and changes tend to appear in continuous bursts, rather than single transactions. Rather than transmit database change records continuously during these bursts, it's often more appropriate to sample that database at intervals, using a tracking algorithm very similar to that of the clock synchronization protocol described above.

- *Database tracking inquiry* message sent by either of a pair of systems, when (1) requested by software changing database (2) check interval timer expires,
- message has form of ICMP timestamp message,
- *identifier* used as a token for specific database,

- Message time fields used for time of last change at sender, current time at sender, and time of last change at responder,
- Local clocks are synchronized using separate protocol, above,
- *Sequence* field indicates number of changes (or similar quantitative measure of change rate, e.g. slew rate),
- Zero change rate in conjunction with *approximate concurrency* of last change (see below) causes increase in interval of check, up to a maximum,
- Nonzero change rate, non-concurrency of last change causes follow-up exchange of *database update* messages (defined separately) and decrease in check interval, down to a minimum.

Because no averaging is needed to judge concurrency, and because synchronization and concurrency of change are established separately, the database tracking protocol is simpler than that for timebase synchronization. It also is quite robust (can survive communication errors) if little state information is held between check intervals. If state information is held in the database, and as a result lost data will corrupt the object which contains the database, a reliable link protocol must be used to carry these messages. In the event of loss of connection or other uncorrectable error, the protocol should request database initialization – i.e. complete update – on resumption of the connection.

The principal protocol parameters are:

- Maximum concurrency check interval,
- Minimum concurrency check interval,
- Interval increment (or factor),
- Delta-t for *approximate concurrency* condition; normally based on clock synchronization precision (based on clock tracking protocol parameters, including network distance to master clock).

In addition, the database update messages are defined separately according to application-specific needs. (A reliable protocol may be necessary.)

```
 0                   1                   2                   3
 0 1 2 3 4 5 6 7 8 9 0 1 2 3 4 5 6 7 8 9 0 1 2 3 4 5 6 7 8 9 0 1
+-+-+-+-+-+-+-+-+-+-+-+-+-+-+-+-+-+-+-+-+-+-+-+-+-+-+-+-+-+-+-+-+
|     Type      |   Code = 2    |          Checksum             |
+-+-+-+-+-+-+-+-+-+-+-+-+-+-+-+-+-+-+-+-+-+-+-+-+-+-+-+-+-+-+-+-+
|        Database Token         |       Change Rate Index       |
+-+-+-+-+-+-+-+-+-+-+-+-+-+-+-+-+-+-+-+-+-+-+-+-+-+-+-+-+-+-+-+-+
|         Originator's Last Database Update Timestamp           |
+-+-+-+-+-+-+-+-+-+-+-+-+-+-+-+-+-+-+-+-+-+-+-+-+-+-+-+-+-+-+-+-+
|                Originator's Current Timestamp                 |
+-+-+-+-+-+-+-+-+-+-+-+-+-+-+-+-+-+-+-+-+-+-+-+-+-+-+-+-+-+-+-+-+
|          Responder's Last Database Update Timestamp           |
+-+-+-+-+-+-+-+-+-+-+-+-+-+-+-+-+-+-+-+-+-+-+-+-+-+-+-+-+-+-+-+-+
```

Fig. 3. Database Tracking Message and Reply

This protocol also can be implemented using the ICMP Timestamp Message structure, identifying the database concurrency protocol with a unique value of Code. Identifier is used as a token for the particular database under consideration, and Sequence becomes the "slew rate" indicator. A maximum value of Sequence can be used to indicate a request for database initialization.

3.4.4 Clip Window Zoom Synchronization

This tracking protocol for database concurrency is especially appropriate for a large class of moving frame-of-reference problems in user interface design at the experimenter's workstation, all based on the principle of deletion of data beyond the "edges" of the display window.

This concept is fundamental to computer graphics, in both the "viewport" of display device screen coordinates, and the "window" into real-world coordinate data that is displayed on it. Without clipping, a display program spends much of its time placing points off-screen; efficient methods of decision on whether a particular point is inside or outside these rectangular windows were found early in the field's development. [Sutherland74]

3.4.4.1 Generalization of Clipping

A more powerful form of clipping described by [Vatti89] is based on a polygon region on the screen used as a clip window, rather than a rectangle. Because the method determines the intersection of two polygons (the clipping and the clipped) it provides a basis for efficient operations on two-dimensional regions (union, intersection, exclusion, and difference). Although Vatti's work was directed toward algorithms implemented in physical device coordinates, it is also applicable to clipping in world coordinates.

3.4.4.2 Data Stream Clipping as a Distributed Operation

These methods can be applied to significant practical problems. In displaying information from observation systems or sensors located at stations over a large geographic area, communications bandwidth to the display is a very precious resource, because typically only low-speed serial interfaces are available. Because only a limited region is of interest and on the display at any time, it is of extreme importance that no bandwidth be consumed by data outside that region, and that data regarding the region and the display characteristics be stored locally within the display station. The operation of suppressing data messages regarding points at world coordinates outside the current display window is then a form of world-coordinate clipping, and the window is a simple distributed database shared by the systems managing the data stream (possibly including the remote sensor) and the experimenter's workstation. To minimize the bandwidth requirements of the system, the user's manipulation of screen objects to change the clipping window should *push down* as far as possible, to throttle unnecessary portions of the data stream close to the source. (The point where the stream is controlled will here be called the *data stream manager*, for want of a better term.)

3.4.4.3 Clipping Window (Region) as a Distributed Database

This clipping down may be a simple rectangle, defined by two extreme corner coordinate pairs; but in a more general user interface, the clipping window may be a *region of interest* with more general polygon form, or perhaps a set of points. Additional conditions can be imposed on the selection of data points from within that region, in the manner of IBIT. A region then consists of a boundary definition (possibly a point set) and an algorithm for point membership. This object should be dynamically alterable by user interaction – through dragging, zooming, rubber-banding of boundaries, etc., and the consequences of this change should include alteration of the data stream into the workstation as well as alteration of the display.

This problem can be solved within limits set by the communications bandwidth available, as a distributed database tracking problem of the type described above. As a region of interest is defined on the screen, its components are created as world-coordinate objects in both workstation and in the data stream manager. As the region of interest is activated, data events within the region are accepted by the data stream manager for transmission upstream; and as it changes in location or boundary, the region of acceptance tracks the region of interest.

3.4.4.4 Protocol for real-time Distributed Clip Region Database Update

This scheme provides a method of panning, zooming, and otherwise manipulating a window in world coordinates, tracking changes in a corresponding window at the workstation. The database consists of a list of world coordinates defining a polygon, and an algorithm selector (identifying use of rectangle corner pair, vertex list, or point set) and a timestamp; it is replicated at workstation and at data stream manager. When the operator begins zooming, the workstation issues a *window update* message containing the full window database, and the data stream manager changes the clip window accordingly. As the clip processing continues with the new window, the data stream manager and display window manager exchange *window tracking* messages, followed by a new *window update* message from workstation to data stream manager on a change. A zoom ends when the window update messages subside, and window tracking messages are exchanged at the maximum interval.

The protocol is composed of:

- an data stream consisting of data records prefixed by a header which is identical to the ICMP Timestamp Message, where Code now is chosen to uniquely identify uncorrected data,
- The clock sync tracking protocol defined above,

```
0                   1                   2                   3
0 1 2 3 4 5 6 7 8 9 0 1 2 3 4 5 6 7 8 9 0 1 2 3 4 5 6 7 8 9 0 1
+-+-+-+-+-+-+-+-+-+-+-+-+-+-+-+-+-+-+-+-+-+-+-+-+-+-+-+-+-+-+-+-+
|     Type      |   Code = 4    |          Checksum             |
+-+-+-+-+-+-+-+-+-+-+-+-+-+-+-+-+-+-+-+-+-+-+-+-+-+-+-+-+-+-+-+-+
|       Database Token          |        Sequence Number        |
+-+-+-+-+-+-+-+-+-+-+-+-+-+-+-+-+-+-+-+-+-+-+-+-+-+-+-+-+-+-+-+-+
|     Originate Timestamp                                       |
+-+-+-+-+-+-+-+-+-+-+-+-+-+-+-+-+-+-+-+-+-+-+-+-+-+-+-+-+-+-+-+-+
|     Receive Timestamp                                         |
+-+-+-+-+-+-+-+-+-+-+-+-+-+-+-+-+-+-+-+-+-+-+-+-+-+-+-+-+-+-+-+-+
|     Transmit Timestamp                                        |
+-+-+-+-+-+-+-+-+-+-+-+-+-+-+-+-+-+-+-+-+-+-+-+-+-+-+-+-+-+-+-+-+
|     Data Record                                     |
+-+-+-+-+-+-+-+-+-+-+-+-+-+-+-+-+-+-+-+-+-+-+-+-+-+-+-+-+-+-+-+-+
```

Fig. 4. Database update message and acknowledgement

- The database tracking protocol defined above,
- A database for region of interest, as defined immediately above,
- A reliable protocol for exchange of updates of this database.

The reliable protocol can also be implemented using a ICMP-structured message header, in which Code uniquely identifies a database update protocol; Identifier contains the token for the database; and Sequence contains a change sequence number for the exchange. A data record follows; the Checksum validates data in the whole message. A database update response ACK message is provided by returning a copy of the header without its data. NACK is provided by returning a maximum value, or other invalid value, in the Sequence field.

4. Example: Display of real-time Lightning Weather Information

The FAA is examining use of real-time lightnight location data provide by two separate networks of lightning locator systems, operated by the National Forest Service east of the Mississippi, and by the Bureau of Land Management west of the Mississippi. The data from the two networks is merged by a system operated by SUNY-Albany, and distributed by sattellite link to research users, including the US DOT Transportation Systems Center in Cambridge MA, where a communications concentrator and distributor process is being evaluated for use on this data. One possible use of this data stream is distribution to remote FAA PC systems dialing in to the FAA concentrator system over low-speed asynchronous connections. The simplest solution to graphic display of lightning data in this configuration is simple graphic terminal emulation at the PC (e.g. VT240) for a straightforward GKS program running at the concentrator. Zoom and pan would be unacceptably slow, because of redrawing of base map information; and in peak periods, the data rates from large regions over the US would far exceed the available communications capacity.

To solve both these problems, the application should be divided into a local PC display manager program, and a data stream manager program at the concentrator. The base map and window control would be the responsibility of the display manager, with as much base map information as is necessary available locally. (A tracking protocol could also offer the ability to add base map information where missing.)

This PC display manager application and the protocols supporting it rely on accurate maintenance of a local clock with resolution greater than one second. Methods of implementing this clock in DOS software on a standard PC are suggested in [Lang89].

5. Other Applications

The methods described above are directly applicable to a wide variety of experimental problems where data collection over a geographic area is used to feed a data integration and display system, and where workstations must be located remotely or perhaps in mobile stations, accessible only over low-speed serial links.

A similar problem is solved in a similar way in the experimental data collection system of IBIT, where sensor data stream are enabled a necessary when called on as condition variables in a spreadsheet-like user interface at the workstation. This indicates a possible larger generalization of the method to systems where discrete clip sets form *regions*.

One such application immediately comes to mind: a remote inquiry system for patient monitoring equipment, providing a doctor with the ability to survey vital signs of a large population of monitored patients, consuming minimum required bandwidth per patient during normal periods, but providing maximum bandwidth during inquiry and examination. The data stream clipping method can support interactive or automatic selection of sets of critical patients for continual monitoring. The technique works equally well, whether the patients are located remotely (e.g. home monitoring of outpatients) or the doctor is located remotely (e.g. at home on emergency call).

6. Summary

The presence of communication protocol subsystems in distributed real-time systems was noted, and the value of formal methods in design implementation of custom protocols for these systems was discussed. Certain common problems in distributed real-time systems seem susceptible to a toolkit approach, where methods and strategies, if not formal protocol standards, can be applied to recurrences of similar problems. A common problem of clock synchronization was addressed with a derivative of a standard TCP/IP Internet Control Message Protocol (ICMP) Timestamp Message as the basis of a tracking algorithm. A similar tracking algorithm and protocol for database concurrency then was developed using this clock synchronization scheme as its basis, also using derivatives of the ICMP Timestamp Message. This database tracking protocol is developed as the foundation of a method of controlling data flow in a distributed data collection system (specifically an experimental FFA weather system) in conjunction with interactive pan and zoom of a region of interest at a PC workstation accessible only through a low-speed serial link. This method of *data stream clipping* permits greatly improved base map plotting during pan and zoom, using local map data within the PC, while limiting the data sent to the PC from the weather information system to only that of specific interest, in real-world coordinates and units. The method suggests generalizations to other applications and problems, indicating that the toolkit approach to custom experimental workstation protocol software can open up new research and development opportunities.

References

[Bartram89] Bartram, L. R., Booth, K. S., Cowan, W. B. Issues in the Design of Workstations for Psychology experimentation. [In this volume.]

[Carson82] Carson, George S. Message-Based Distributed Computing. IEEE 1982 Real-Time Systems Symposium (1982), pp. 170–183

[Chesson88] Chesson, G. XTP Definition, Revision 3.1. Protocol Engines Inc., March 28 1988

[Cohn88] Cohn, Marc. A lightweight transfer protocol for the US Navy SAFENET LAN standard. IEEE 13th Conference on Local Computer Networks (1988)

[Comer88] Comer, D. Internetworking with TCP/IP: Principles, Protocols, and Architecture. Prentice-Hall, 1988

[DaCruz87] DaCruz, Frank. KERMIT: A File Transfer Protocol. Digital Press, 1987

[Dijkstra68] Dijkstra, E. W. Cooperating Sequential Processes. In Programming Languages, F. Genuys (ed.), Academic Press, New York, 1968

[Gentleman8] Gentleman, W. Morven. Message Passing Between Sequential Processes: the Reply Primitive and the Administrator Concept. Software – Practice and Experience 11, 435–466 (1981)

[Gilbert72] Gilbert, Philip, and Chandler, W. J. Interference Between Communicating Parallel Processes. CACM 15(16), 427–437 (1972)

[Halsall88] Halsall, Fred. Introduction to Data Communications and Computer Networks. Addison-Wesley, 1985

[Hansen78] Hansen, P. Brinch. Distributed processes: a concurrent programming concept. CACM 21(11), 934–940 (1978)

[Hoare78] Hoare, C. A. R. Communicating Sequential Processes. CACM 21(8), 666–677 (1978)

[IEEE83] IEEE. Oplen Systems Interconnection (OSI) - New International Standard Architecture and Protocols for Distributed Information Systems. (Special issue) Proceedings of the IEEE, 71(12)

[ISO7498] International Standards Organization. ISO Open Systems Interconnection – Basic Reference Model

[Kaminski86] Kaminski, M. Protocols for communicating in the factory. IEEE Spectrum, April 1986

[Koegel89] Koegel, J., McElroy, S., and Chanasyk, B. Data Analysis in a User Interface for Distributed Process Control. [In this volume.]

[Lang89] Lang, Phyllis K. Improving the Timing Resolution of an MS-DOS PC. C Users' Journal, July 1989, pp. 85–91

[Lesk75] Lesk, M. E., and Schmidt, E. Lex – A Lexical Analyzer Generator. AT&T Bell Laboratories Technical Report 39 (UNIX), 1975

[Mac87] Apple Computer Inc. Technical Introduction to the Macintosh Family. Addison-Wesley, 1987

[O'Reilly88] O'Reilly, T., Quercia, V., Lamb, L. X Window System User's Guide. O'Reilly & Assoc., 1988

[Shaw86] Shaw, Alan C. Software clocks, concurrent programming, and slice-based scheduling. IEEE 1986 Real-Time Systems Symposium (1986), pp. 14–18

[Stallings88] Stallings, W. Handbook of Computer Communications Standards (3 vols). Howard Sams Co., 1987

[Sutherland74] Sutherland, Ivan E., and Hodgman, Gary W. Reentrant Polygon Clipping. CACM 17(1) 32–42 (1974)

[Tanenbaum88] Tanenbaum, Andrew. Computer Networks, 2nd ed. Prentice Hall, 1988

[Vatti89] Vatti, Bala R. A Generic Solution to Polygon Clipping. (Personal communication, 1989)

[Volkman89] Volkman, Victor. MKS Lex and Yacc – A User's View. C Users' Journal, July 1989, p. 77ff.

Data Analysis in a User Interface for Distributed Process Control*

J. Koegel, S. McElroy, B. Chanasyk

Abstract

IBIT is a user interface toolkit for distributed process control and manufacturing applications. It has been used to build a direct manipulation interface for a family of ion beam implanters. In the course of controlling production, a large volume of information can be collected for both online and offline analysis. In this paper we discuss extensions to the toolkit related to data collection and analysis. These extensions, which are currently under development, include a spreadsheet for event definition, icons for controlling charting, and visualization of the ion beam density. We relate these results to data analysis in manufacturing applications in general, and to facilities needed in workstations for experimentation.

1. Introduction

We have developed an object-oriented user interface toolkit (IBIT) with specific features for applications in distributed process control and manufacturing. This toolkit has been used to construct a direct manipulation user interface for a distributed control system used to control a family of ion beam implanters. This system, the Eaton NV-20A ion beam implanter, is currently installed in the field and used in production environments. A number of extensions to the user interface toolkit are currently being developed. This paper addresses the addition of facilities for data analysis.

The particular equipment being operated has several hundred I/O points, many of which have sampling times in the microseconds. A distributed control system (Fig. 1) has been designed which consists of a hierarchy of processors which are networked together. The user interface processes reside on the color graphics

---

*All work described in this paper was done as part of the Eaton Corp., Semiconductor Euipment Division, development of the NV-20A ion beam implanter.

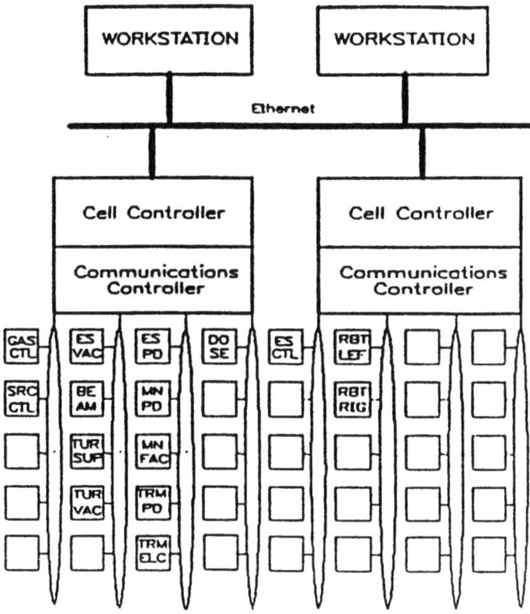

Fig. 1. NV-20A distributed control system

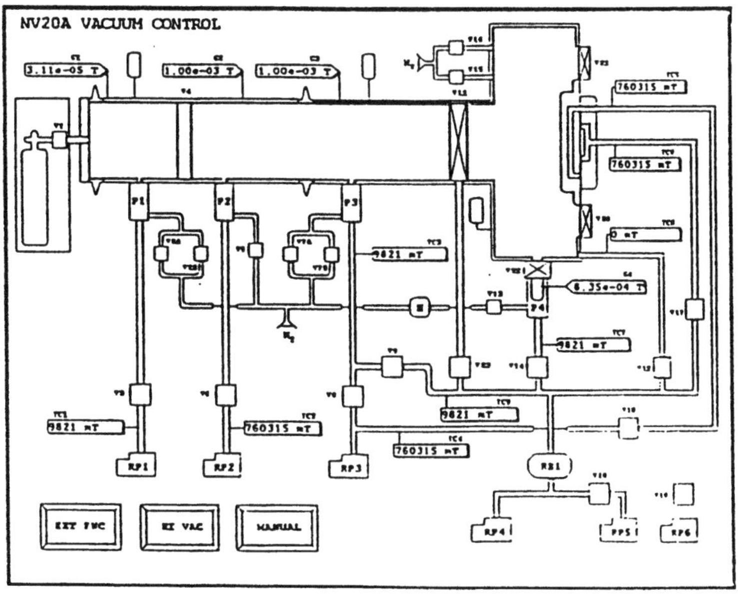

Fig. 2. NV-20A vacuum control screen

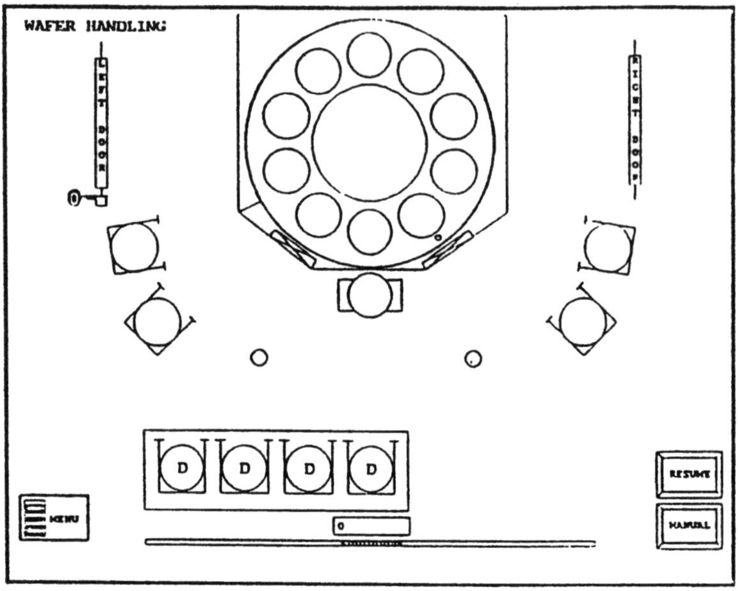

Fig. 3. NV-20A robot end station screen

workstation and, following conventional practice, contain no application state or control algorithms. The user interface allows the operator to examine the state of the machine and to send commands to various control processes.

Because of the complexity of the machine, the user interface provides individual screens for specific subsystems, such as the vacuum system (Fig. 2) or the robot station (Fig. 3). Additionally, two status panels are always available. The lower panel displays alarm and status information; the panel on the right displays the control state of the machine. The operator is free to shift between screens as needed. However, depending on the state of the machine, certain operations are disabled.

Tools for data collection and analysis are needed for such tasks as production quality monitoring, re-running as process at specific settings, and for scheduling maintenance. Historically, data collection and analysis were done offline on a remote host. While this mode of analysis will remain, a direct manipulation interface on a graphic workstation offers the possibility of online analysis, charting, and logging.

The remainder of this paper deals with user interface support for data analysis in a distributed process control application. The next section examines the requirements by discussion of representative examples taken from the NV-20A ion beam implanter. This is followed by a description of the salient features of the toolkit. The subsequent two sections present the design currently being pursued which incorporates icons and direct manipulation techniques. The final sections summarize the state of the implementation and related work.

2. Data Collection and Analysis Requirements

The NV-20A is a medium/high current ion beam implanter used to introduce ions into silicon wafers. The basic operation of the machine is as follows:

1. load the set of wafers to be implanted,
2. create a high vacuum in the chamber containing the wafers,
3. bring the ion beam up and "tune" it for the desired behavior,
4. perform an implant operation,
5. turn the beam off, return the chamber to atmospheric pressure and remove the wafers.

While the basic cycle of operations is repeated over and over, actual operating parameters can vary significantly from run to run. Several important aspects of this production process are:

1. The machines have operating characteristics that are difficult to repeat on successive implants, and which also vary from machine to machine
2. The tuning process is crucial to the success of an implant, and although semiautomated, is highly interactive and can be enhanced by online charting facilities
3. Recorded tuning variables can be used in subsequent implants.

2.1 Data Collection

As mentioned previously, there are hundreds of I/O points which might be of interest to users of the machine. In some instances, the data is collected for short term use, for example, to analyze a single production run. In other cases the data is collected to maintain an historical record, for example, to evaluate component wear or to schedule maintenance. The values of a given point might be of interest only when the machine is in a specific state; the identification of the state might involve identifying states of specific subsystems. The capabilities needed for data collection include:

1. Selection over a large number of possible points
2. Specification of sampling rates by time or frequency
3. Sample triggers defined by changes in machine state, changes in point values, point values exceeding a specified range, or some combination of these
4. Specification of sample windows by triggers or by elapsed time

In section four we present an interactive facility for defining data collection events. This facility uses the direct manipulation interface to the machine as a spreadsheet-style method for recording the event algebra.

2.2 Data Analysis

The data analysis facilities envisioned for the NV-20A include:

1. plot multiple variables on same graph,
2. plot several historical samples on same graph,
3. interactively select and deselect points to be sampled and events which control the sampling,
4. comparison of a set of variables across several machines.

More advanced features (filtering, statistical analysis, FFT or other transforms, extrapolation, interpolation) could easily be added to this list depending on the application. The icon-based tools described in section 4, together with the event definition facility, are sufficiently general to support both the basic and advanced analysis capabilities.

2.3 Examples

The charting facilities need to support a diverse set of graphs. Examples from the NV-20A include:

- AMU (Atomic Mass Units) Spectrum,
- Beam tuning settings (about a dozen variables),
- Wafer eccentricity plot used for alignment of wafers,
- Temperature variation of sonic sensor used at end station.

Mackinlay [Mackinlay86] describes techniques for automatically generating graphical presentations with regard to criteria of expressiveness and effectiveness. These techniques can be applied to the analysis facility to enhance the informativeness of the charts.

3. Overview of the IBIT Toolkit

The interface is composed of a set of screens which partition the overall operation of the machine into operational sub-systems. Each screen is composed of a set of objects which represent machine components (e.g. gas valves, vacuum chambers, silicon wafers, wafer cassettes) whose state is dynamic and with which the operator may interact in some way.

Objects are organized by type. Complex objects can be constructed by the aggregation of other objects. Dynamic behavior of objects is mediated by object methods which respond to changes in the machine I/O points associated with the object or to events at the interface.

In these respects IBIT is similar to other object-oriented user interface toolkits that have been described elsewhere. Additional features have been introduced which provide important support for distributed process control applications:

- *security features:* in a given environment, different operators may have different privileges affecting their ability to control the screen; the interface must enforce these privileges is an easily administered and understood manner
- *alarm conditions:* control processes detecting unusual machine conditions notify the operator by posting an alarm; alarms are given different severity levels and are logged for later review
- *performance issues:* rapid update of the dynamic state of display objects with respect to the I/O points is crucial to the effectiveness of the interface; the large number of points and their distributed nature requires specific design in order to meet the performance goal

3.1 Interface to Control System I/O Points

Objects may be associated with one or more data points which are managed by low-level control systems. These data points are updated by a data acquisition process running asynchronous to the display process. The acquisition process and the display process communicate via shared memory. Whenever a screen associated with a particular sub-system is displayed, the display process informs the data acquisition process of the set of data points which are specific to that sub-system. The data acquisition process fills in the slots in shared memory with the data it receives and the display process updates the graphical representation of the objects under its control.

3.2 Event Handling

Objects can be made to respond to an arbitrary set of events. Typically they respond to update events generated by the display manager. Update events are the mechanism which keeps the display graphics in sync with the data points acquired from the control system. Events also exist to respond to any of a number of operator input events (mouse movements, button clicks, keyboard input). In this sense the objects can be seen as state machines. As the IBIT objects react to events, attributes of the objects are modified as necessary. Among these attributes is the next set of events for which an object will respond. Further, event processing attribution can propagate through the hierarchy of objects.

4. Extensions for Data Acquisition

The features of IBIT discussed so far were designed and implemented to provide the human interface for operating the machine in a production environment. IBIT can also support an environment more useful for scientific analysis. To do this the data acquisition concepts presented thus far need to be extended. First the data needs to be more persistent than the data used to keep the process control display objects up to date. Second, it is also desirable to limit the quantity of data to that which is actually useful in the analysis. Finally the package must support mechanisms for displaying the data in various formats more useful for analysis.

4.1 Event Definition

IBIT provides a spreadsheet form of event definition (Fig. 4) for the purpose of establishing when data sampling is appropriate. The screen is divided into two areas. The bottom represents a spreadsheet where each cell has a logical true/false value. These values are derived by evaluating the grammar described below. The cells are ANDed together in a row-wise fashion to form a complex logical expression which is called an event definition. The event definition can be evaluated to trigger an event. The event definition is constructed from expressions in the event language, the syntax of which is described next.

4.1.1 Event Language

The following grammar defines the language recognized by the event definition spreadsheet. This syntax allows expressions involving I/O points and machine states to be combined to form an event definition. An event definition names a row in the spreadsheet containing one or more event expressions. The mapping from event definition to the corresponding set of event expressions in the specified row is given in Appendix A. An event definition is named by the "event" keyword and the row number where the definition is entered, as in "event 3".

<event_definition> → *event* *<row_id>*

An event expression can be one of 1) a relational expression involving I/O points, 2) a temporal expression involving machine states, 3) a reference to an expression defined elsewhere in the table, or 4) empty.

<event_exp> → *<point_exp><rel_op><point_exp>* ||
<state_exp> || *<table_index>* ||
<absolute_tod> || *NIL*

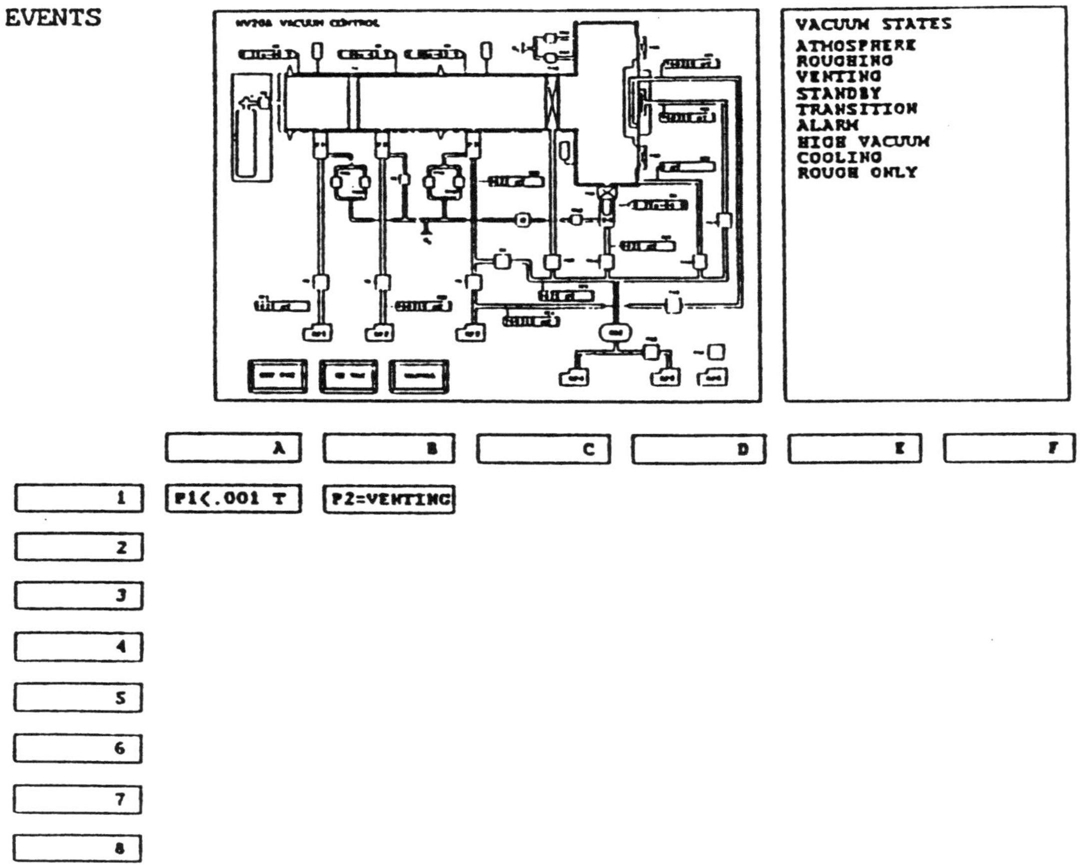

Fig. 4. Event definition screen

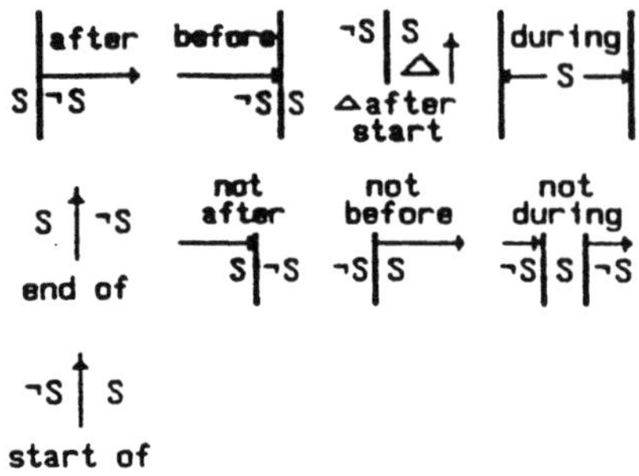

Fig. 5. Examples of event interval and instance icons

Arbitrary arithmetic expressions involving variables corresponding to I/O points are allowed. This portion of the grammar omits the details needed to insure matching units in expressions.

<point_exp> → <simple_point_exp> ||
<point_exp> <op> <point_exp>
<simple_point_exp> → <point> || <value>
<point> → <variable_name>
*<variable_name> → <alpha><alphanumeric>**
<value> → <number><units>
<units> → {Torr, Volt, Amp, ...} || NIL
*<op> → + || - || * || / || ^*
<rel_op> → < || > || <= || >= || == || ≠

All machine states are given predefined identifier labels which can be used to form temporal relational expressions. The interval relations used here are a subset of those found in [Allen82]. The meanings of all the relationships follow the normal sense and are formalized in Appendix A. Figure 5 shows iconic representations used with the spreadsheet. These icons have been designed following some of the principles in [Gittins1986].

<state_exp> → <state> || <interval_relation_1><state><number> ||
<interval_relation_2><state><number> ||
<instance_relation><state><number> ||
<cont_relation> <number>
<interval_relation_1> → {during, after, before} ||
not {during, after, before}
<interval_relation_2> → from <instant_relation> ||
until <instant_relation> ||
between <instant_relation> <instant_relation>
<instant_relation> → {start of, end of, Δ after
start of, Δ before start of, Δ after
end of, Δ before end of}
<cont_relation> → {continuously at Δ intervals}

<state> → <state_identifier_label> || not <state_identifier_label>
<absolute_tod> → time-of-day

Next the table (corresponding to the event defintion spreadsheet) which contains the event expressions is defined. Let

$$row\_id \equiv \{m..n\},\ n,\ m \in Z^{+},\ m < n$$

$$col\_id \equiv \{A..Z\}^{2}$$

$$R = \|\ row\_id\ \|\ ,\ C = \|col\_id\|$$

Then T is a table of dimension $R \times C$, where $T[i,j] \in$ {event expression}, $i \in row\_id$, $j \in col\_id$, and event expression is as defined above. Further, $T_{row}[r] \equiv \{T[i, j]$ where $i = r \in row\_id$ and $j \in col\_id\}$, that is, $T_{row}[r]$ is the set of all event expressions contained in the table at row r.

Let $table\_index \equiv (i,j)$, $i \in row\_id$ and $j \in col\_id$, then the following definitions allow the event language grammar to recognize table references:

<row_id> → row_id
<table_index> → table_index

4.1.2 Examples

Examples of valid event expressions include:

0.1 msec after start of implant active
T1 > 800 °C
V2 < V4 + 100 V
continuously at 1 sec intervals
until .2 sec after start of implant idle
not before roughing
from 1 sec before end of hivac

The user creates an event *definition* by entering one or more event expressions in a given row in the table, one expression per position. The value of an event definition at any given time t is the boolean AND of all the event expression contained in the row, each evaluated at time t. Thus in the event definition spreadsheet, the horizontal direction represents a conjunction of event expressions and the vertical direction represents a disjunction of event definitions.

To simplify entry, the user can select an I/O point by pointing at its representation on one of the subscreens, as shown in Figure 4. The state expressions can be entered using the icons shown in Figure 5. Once an event definition is entered, it is used in any charting or logging operation by reference to the row number, for example, "event 3".

4.1.3 Limitations

The event language does not describe counter-based relationships, such as "the 13th occurrence of the start of implant active". An event counter icon could be introduced as one of the analysis icons described in the next section. When placed in a stream, such an event counter icon would be analogous to a divide-by-n counter used in digital logic.

Repetitive sequence such as "whenever event 1 followed by event 2 followed by event 3 occurs" are also not representable. These could be supported by changing the semantics of successive columns in a row from being order-independent to order dependent, perhaps by introducing the keywords *seq* and *end_seq* to represent the beginning and end of a sequence definition.

4.2 Icon Controlled Charting and Logging

Once the user creates the necessary event definitions as described in the last section, an arbitrary I/O point can be charted and/or logged by use of icons. An instance of the chart icon is attached to the displayed machine point to cause a chart of that variable to be shown. Enabling of one or more event definitions activates the data stream needed for the chart. The parameters of the chart assume defaults that are appropriate for the data being charted (see [Mackinlay86] for results in this area), and can be overruled by the user through interactive means. Multiple points can be charted on separate graphs by using one of the original set of chart icons, or on the same graph by sharing the icon for a previous variable on the common graph. This sharing of icons is accomplished by copying the icon from a previous variable to the new one. In a similar way, a data stream can be selected for logging, replacing the use of the chart icon with the logging icon.

The use of icons allows more sophisticated analysis techniques such as filters or transforms to be used Hirakawa et al [Hirakawa1988] show how a series of icons can be connected or overlapped to represent a sequence of operations on a data stream. This type of iconic programming can be used here by providing icons for the desired analysis operations. The user can connect the icons together for the desired results, relying on menus to enter any additional parameters that may be needed.

4.3 Visualization of Beam Density

Figure 6 shows a time scan and a reconstructed contour map for an ion beam on an earlier implanter. The uniformity of the beam density can be visually gauged using the contour map. In general a uniform beam represents a higher quality tuning of the beam control parameters because the resulting implant has fewer imperfections. The utility of the contour map can be enhanced by a color coding to identify the contour levels.

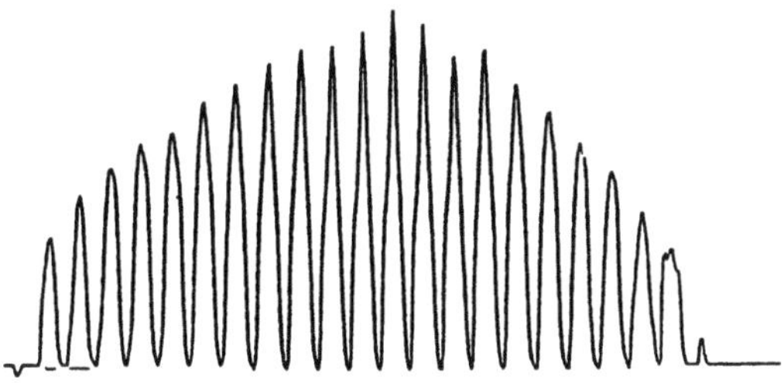

Fig. 6(a). Ion beam density time scan

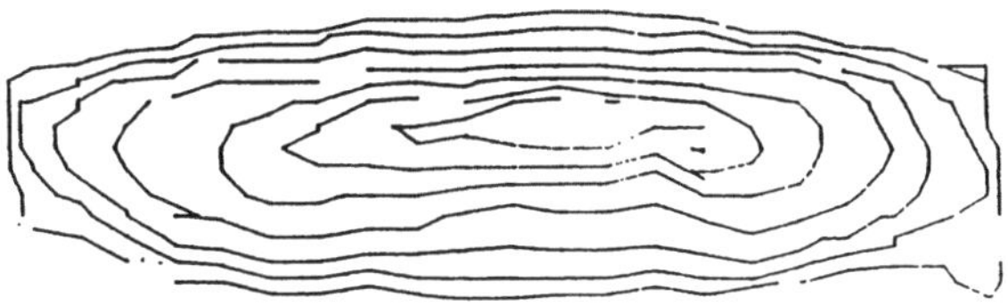

Fig. 6(b). Ion beam density contour map

As mentioned earlier, the beam tuning is a complex operation involving more than a dozen machine parameters. The control path used to reach the tuned state influences the tuning operations as well. Thus tuning of the beam for a given process is not repeatable by simply setting the parameters to the values last used for the process. Because of this and the fact that the implanter is generally used for many different processes which have correspondingly different tuning requirements, beam tuning is difficult to optimize. Software for automatically tuning the beam generally relies on heuristics as well as an analytical model of the machine.

The beam density and the tuning parametes form a data set for analysis. This multidimensional data set is a candidate for more sophisticated visualization techniques than the contour map, which is limited to 3 or 4 dimensions.

5. Application to Manufacturing Data Analysis

Automated manufacturing systems involving process control generally produce a large amount of control and sensor data, the analysis of which depends in part on temporal relationships. The event definition screen, the event language, and the icon-based analysis are well-suited to this type of application. In manufacturing applications where temporal relationships are less significant (e.g., light metal-working, assembly), the event facility described is probably less appropriate. The icon-based analysis technique could be used for analysis of data from part inspection and quality.

6. Application to Workstations for Experimentation

Here are two examples where the IBIT event definition facility would be applicable to workstations used for experimentation.

1. A direct manipulation interface used to control a laboratory for experiments
   The laboratory sensors can be selected interactively using the direct manipulation interface and data collected according to interactively specified events.
2. Graphic simulation of some physical process
   Assume that the graphic simulation is displayed so that parameters of interest can be interactively selected. If the simulation's variables (time clock and process values) can be identified and accessed, then the event facility and icon-based analysis can be used to allow the researcher to easily shift focus during simulation runs.

8. Summary

We have presented a set of interface techniques being developed for a specific distributed process application for which a direct manipulation interface already exists. The extensions to the object-oriented IBIT toolkit include an event definition facility and icon-based charting/logging. Rudimentary visualization of the beam density is achievable with the current toolkit, with more sophisticated visualization techniques waiting to be explored. The techniques described here have general applicability to process control manufacturing and experimentation.

References

[Allen83] Allen, J. F. Maintaining Knowledge About Temporal Intervals. *CACM* **26**(11), pp. 832–843

[Gittins1986] Gittins, D. Icon-based human-computer interaction. *International Journal of Man-Machine Studies* **24**, pp. 519–543 (1986)

[Hirakawa88] Hirakawa, M., Iwata, S., Tahara, Y., Tanaka, M., and Ichikawa, T. A Framework for Construction of Icon Systems. *IEEE 1988 Workshop on Visual Programming* (1988), pp. 70–77

[Mackinlay86] Mackinlay, J. Automating the Design of Graphical Presentations of Relational Information. *ACM Transactions of Graphics* **5**(2) (April 1986), pp. 110–141

[McCormick87] McCormick, B. H., DeFanti, T. A., and Brown, M. D. (editors), Visualization in Scientific Computing, *Computer Graphics* **21**(6) (November 1987)

Appendix A

Semantics of Event Language

The meaning (true or false) of an event definition is based on the evaluation of the component event expressions using the evaluation functions defined in the following tables, according to the time at which these expressions are evaluated.

Table A-1. Evaluation of Interval Relations

Interval Relation	*Evaluation Function*
during	*during*(S, t) = true if state is S at time t, false otherwise
after	*after*(S, t) = true if a transition from state S has occurred and no subsequent transition to state S has occurred at time t false otherwise
before	*before*(S, t) = true if no transition to state S has occurred at time t, false otherwise
not expr	true if the evaluation function for expr is false at time t, false otherwise, where expr is one of during, after, or before
from inst	true if the instant *inst* has previously evaluated to true, false otherwise, using the instant evaluation functions in Table A-2 and appropriate values for S, t, and Δ
until inst	true if the instant *inst* has not previously evaluated to true, false otherwise, using the instant evaluation functions in Table A-2 and appropriate values for S, t, and Δ
between *inst1 inst2*	true if the instant *inst1* has previously evaluated to true and the instant *inst2* has not subsequently evaluated to true, false otherwise, using the instant evaluation functions in Table A-2 and appropriate values for S, t, and Δ

Table A-2 Evaluation of Instant and Periodic Relations

Instant Relation	*Evaluation Function*
start of	*start_of*$(S, t, 0)$ = true if a transition to S has occurred since time $t-\varepsilon$, false otherwise
Δ after start of	*start_of*(S, t, Δ) = true if a transition to S has occurred since time $t+\Delta-\varepsilon$, false otherwise
Δ before start of	*start_of*$(S, t, -\Delta)$ = true if a transition to S has occurred an interval Δ before time $t-\varepsilon$, false otherwise; (requires buffering).
end of	*end_of*$(S, t, 0)$ = true if a transition from S has occurred since time $t-\varepsilon$, false otherwise
Δ after end of	*end_of*(S, t, Δ) = true if a transition from S has occurred since time $t+\Delta-\varepsilon$, false otherwise
Δ before end of	*end_of*$(S, t, -\Delta)$ = true if a transition from S has occurred an interval Δ before time $t-\varepsilon$, false otherwise; (requires buffering).
continuoulsy at Δ intervals	*continuously*(S, t, Δ) = true if $t = n * \Delta$, $n \in Z$, false otherwise

Table A-3 Evaluation of Event Expressions

Event Expression	*Evaluation Function*
NIL	true
<table_index>	true if the event expression T[*table_index*] is true at time t, false otherwise
<state_exp>	true if corresponding evaluation function from Table A-1 or Table A-2 is true at time t, false otherwise
<point_exp><rel_op><point_exp>	true if, using *<op>* and *<rel_op>* over real numbers and with appropriate scaling for unit conversion and where *<point_exp>* gives value of corresponding i/o point in database, the expression is true, false otherwise

Given the event expression evaluation function f_c in table A-3, the evaluation of event definition e_r at row r at time t is:

$F(e_r, t)$ = true if, for all E,

$f_c(E, t)$ = true and $E \in T_{row}[r]$ and $T_{row}[r] \neq \{NIL\}$

otherwise $F(e_r, t)$ = false.

This assumes a system clock of resolution ε where $\varepsilon \ll \Delta$ where Δ is the smallest interval of interest.

Scientific Visualization in Supercomputing Environments

G. Fabiani, M. Lanzarini, and L. Moltedo

Abstract

This paper describes the CINECA supercomputing environment in which many graphical applications are being realized using local network facilities. The general guidelines of a CNR project on scientific visualization promoted with the cooperation among several Italian research groups are mentioned here. Moreover, the applications considered in this project are provided with graphical workstations remotely connected to a supercomputer via a high speed network.

Introduction

Two recently developed technologies, the supercomputer and the graphical workstation, have essentially changed scientific research and the ease with which the whole computing process can be executed.

The supercomputer allows researchers to address questions and find solutions that could not previously be obtained, offering unequaled computational capabilities. Meanwhile, workstations provide user friendliness, software and productivity tools that contribute significantly to the enhancement of the user's possibilities. By associating the supercomputer with graphical workstations, we can optimize the whole scientific production cycle.

Scientific Visualization deals with the transformation of data from a symbolic representation into a geometric representation, thus enabling researchers "to observe" their simulations and computations. Visualization includes both image understanding and image synthesis; i.e., it is a tool which is set to interpret image data fed into a computer, and to generate images from complex multi-dimensional data sets [1].

However the visualization process demands new standards for the construction and diffusion of visual information. The near future holds faster networks and more distributed computing. It requires new software, hardware and networking standards. In fact, researchers must be able to share model data and digital images among different environments and there is a need for some degree of standardization.

We also have to point out that interactivity and animation represent the main qualities of a visualization. Large-scale supercomputer simulations have traditionally been performed using long runs in batch mode, but at the moment there is a definite trend towards a more interactive use of supercomputers, with the researcher working "on-line" with the computational loop. For instance, this makes it possible to guide a simulation, to explore a parameter space and use high-bandwidth visual output to give back the results. The scientist would like to monitor the progress of a long run in order to more effectively manage the use of expensive supercomputer time. This would best be done by a quick-look visualization capability, which again implies a requirement for high-bandwith graphics.

Since scientific visualization is a modern discipline that is becoming absolutely necessary to let the scientist properly increase his knowledge in a computational laboratory and since the supercomputing centres are so few, it is necessary to design and manage data communication networks that enable remote scientific visualizations.

In 1988 CINECA (Centro di Calcolo Interuniversitario dell' Italia Nord-orientale), the largest Italian Scientific Data Processing Center, designed a new communication network mainly to speed data transmission up to 2 Mbp/s either by land or by satellite. CINECA designed this network considering the need of interconnecting a large number of workstations of different vendors dedicated to different applications.

In 1989 CNR (Italian National Research Council) became a part of this project, giving rise to an initiative aiming at the evaluation of the effectiveness and the fruitfulness of such remote service on the basis of application samples.

In the first section of this paper, we describe the local computing environment and the network facilities of CINECA. In particular, the characteristics of ViScLab (Laboratory of Scientific Visualization) are emphasized.

In the second section, some scientific visualizations which have been realized at ViScLab are pointed out.

In the last section, we describe the aims of a National research effort, partially sponsored by CNR, within the Finalized Project named "Information Systems and Parallel Computation".

This research is addressed to experiment graphical workstations connected to a supercomputer by a high-speed network. We also mention some application fields which we intend to investigate.

1. CINECA

CINECA, created in 1969 by decree of the President of the Italian Republic, is a Consortium which includes 13 Italian Universities and is dedicated to both Governmental and Industrial Research. Primary goals of CINECA are to provide the Italian research Community with the most advanced hardware and software tools for data processing, to promote their use and to develop the necessary know-how.

Since its beginning, CINECA has been the Italian computing centre dedicated to large scale computation thanks to the availability of supercomputers it provides.

The current computing system (Fig. 1) consists of three functional levels:

1. The communication network level, which is able to handle many different protocols, the "old" SNA, DECNET, X.25 and the "new" TCP/IP;
2. The front-end system level which includes a CDC CYBER 180/860 A, two VAXes, a VAX 8800 and a VAX 11/785 and a vector mainframe, an IBM 3090-60E VF, with 6 processors and 6 vector facilities;

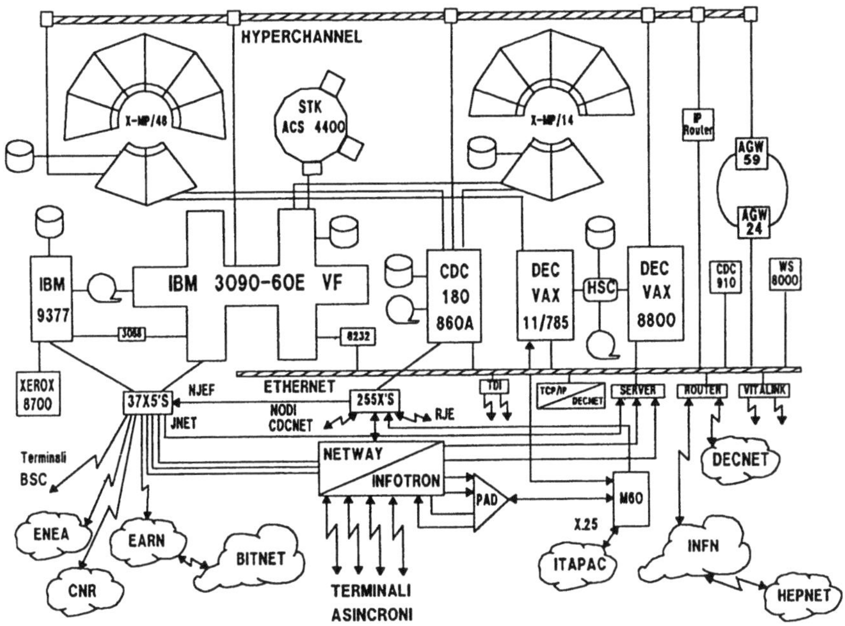

Fig. 1. The current (May '89) configuration of CINECA computing system

3. The supercomputer level which, at the moment, includes a 4 processor CRAY X-MP/48 and a single processor CRAY X-MP/14, which are about to be replaced by a next generation supercomputer, a CRAY Y-MP8/432, that initially will have 4 processors and 32 Mwords of main memory.

1.1 CINECAnet

The CINECA communication network, named CINECAnet, is a typical star network connecting 22 different cities to the CINECA computing centre via dedicated lines with speed ranging from 9600 bps up to 2 Mbps (Fig. 2).

In this paper we focus our attention on the fastest connections which are implemented either by satellite or coaxial cables, depending on the distance (Fig. 3).

The current satellite sub-network consists of three remote stations: CISCA (Centro Interdipartimentale Servizi per il Calcolo Automatico) in Trento, SISSA (Scuola Internazionale Superiore di Studi Avanzati) in Trieste and CIRA (Centro Italiano Ricerche Aerospaziali) in Napoli. The key feature of this network is to be found in the implementation of a double speed scheme using a single channel per carrier (SCPC) mechanism at 64 Kbps when a remote site transmits data to the central site and using a broadcast mechanism of a common 256 Kbps channel when the central returns data to the remote. This scheme is particularly suitable for remote graphics applications where "few" input data can generate a lot of output results.

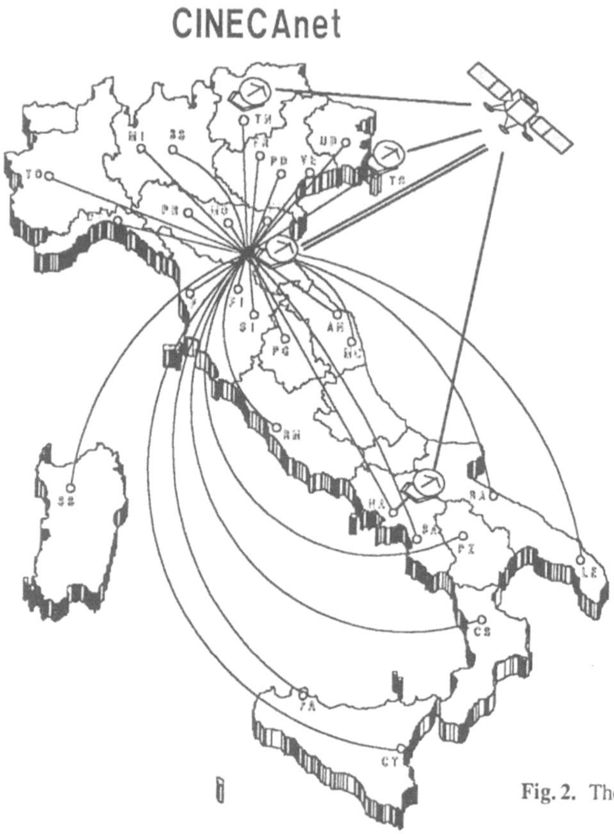

Fig. 2. The star network of CINECA

The coaxial sub-net will have a 2 Mbps speed to connect CINECA and two remote sites at the University of Bologna with a triangle topology.

CINECAnet is a collection of four separate sub-networks each one with a different architecture and protocol: START/STOP, DECNET, SNA and TCP/IP.

START/STOP uses low speed channels, DECNET and SNA sub-networks use medium speed channels and TCP/IP is implemented in the high speed network using satellite and coaxial cables as transmission media and includes supercomputers, mainframes and workstations.

TCP/IP was selected because of its ability to run either on Ethernet (10 Mbp/s) or communication lines (9600 bp/s – 2 Mbp/s) for almost all the workstations which are available on the market and almost all the proprietary operating systems. In this way, CINECAnet can connect remote Ethernets to the purpose of a better integration of computing resources, thus also increasing connectivity and transparency between users and systems.

If this new computing environment with the increased capability of supercomputers and the increased speed of the communication lines is actually going to be widely used by Italian scientists, then the availability of significant graphics facilities turns out to be a necessary goal to achieve.

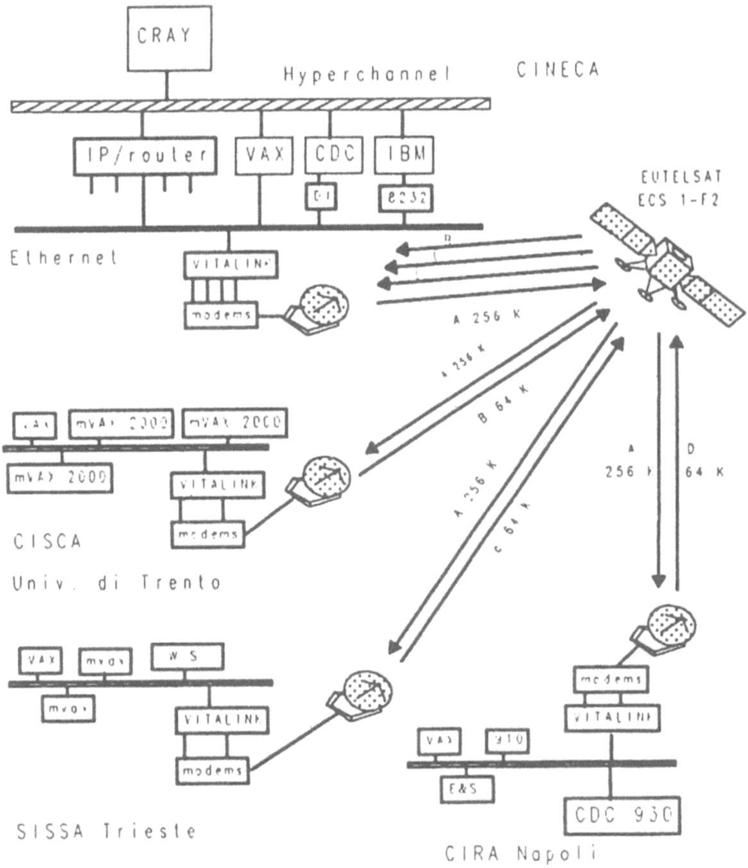

Fig. 3. Integration of LAN via satellite

1.2 ViScLab, a Laboratory for Scientific Visualization

In its computing centre CINECA has developed a Laboratory for Scientific Visualization (ViScLab) [2], whose main aim is to provide the whole scientific Italian community with interactive visualizations of their computational results.

ViScLab includes (Fig. 4), a number of TEKTRONIX 4111, graphic devices and by the following workstations: APOLLO DN 590T, APOLLO DN 4000, IRIS 4D/80GTB, CDC910B (equivalent to an IRIS 3130), DEC WS8000, IBM 5080.

All the workstations are connected by an ETHERNET local network via TCP/IP. Furthermore, they are connected by Hyperchannel (50 MBytes/sec) to the CRAYs to allow interactive access to the supercomputers.

ViScLab also includes devices to permit image acquisition (by Sony DXC 3000 camera) and the recording of image sequences both on videotape (VTR Sony BVU950P 3/4″ PAL), and 35-mm film (QCR-MATRIX).

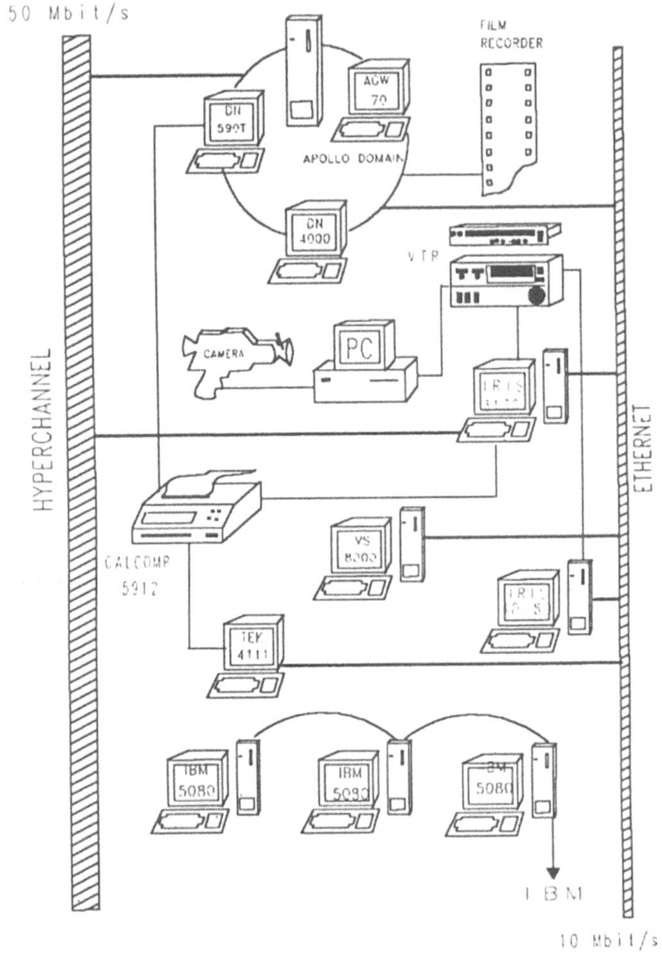

Fig. 4. The configuration of the Laboratory for Scientific Visualization (ViScLab) at CINECA

It should also be pointed out that this system allows the recording of frames produced by any component of the computing environment.

The graphical packages which are available at CINECA are evolving more and more in order to satisfy both the user's needs and technological progress. At the moment they include: GKS, PHIGS, DISSPLA, UNIRAS, CSADIE, MOVIE.BYU, OASIS.

Some of these packages are available both on supercomputers and different workstations. For instance, the same GKS implementation is available on CRAY and VAX but on APOLLO workstations there is a proprietary GKS. Likewise different PHIGS implementations are obviously available on DEC and APOLLO workstations.

Since one of the approaches CINECA clearly tends to choose is the realization of "de facto" portable applications, a ViScLab group is now carrying out interpreters which will allow to hide the differences among the various implementations of a same standard proposal, which, though slight in themselves, still exist.

2. The CINECA Local Approach

In the introduction, we point out that interactivity and animation are relevant techniques in scientific visualization. In fact, the researcher can interact with his application so that he can simultaneously analyze physical phenomena and evaluate numerical results. The dynamical visualization provides a deep understanding of the phenomena much better than the traditional analysis of tables and figures. Furthermore, transparency, shading, multiple light sources, reflections, refractions and so on increase the information given back by solid images substantially.

We should also point out that the animation of images produces large amounts of data that have to be stored and transmitted.

CINECA provides the researcher with the techniques we have been mentioning so far, in its comprehensive computing environment based on emerging technologies.

In the CINECA approach the promotion of the use of these new technologies is based on multi-disciplinary teams. The scientist, qualified in a specific application research field, cooperates with the supercomputing skilled people, numerical analyst and computer graphics experts.

Some of the applications which have been recently developed with the cooperation of ViScLab staff come from different fields, such as atmospheric physics, astronomy, biomedicine, structural analysis and restoration of artistic works and monuments [3].

Figure 5 concerns a graphical result of a research whose main aim is addressed to the study of an oceanic circulation model carried out by IMGA (Istituto per lo studio delle Metodologie Geofisiche e Ambientali) in Modena.

Figure 6 is a frame of an animation realized in research made by Dipartimento di Astronomia in Bologna set to numerically simulate the interaction of a supernova with the surrounding environment.

Figure 7 represents a three dimensional model of a collagene fibril developed by the Istituto di Anatomia Umana Normale in Bologna for studying interactions among components of the extracellular matrix.

Figure 8 visualizes a three dimensional model of a titanium prosthesis of a femoral stem realized by Sperri S.p.A. in Pordenone in cooperation with CINECA.

All these visualizations are mainly obtained by means of post-processing of results coming from batch computations. Actually, ViScLab team is developing tools to obtain visualization along with computation. This technique is going to be more and more important in the future environment of remote graphical applications.

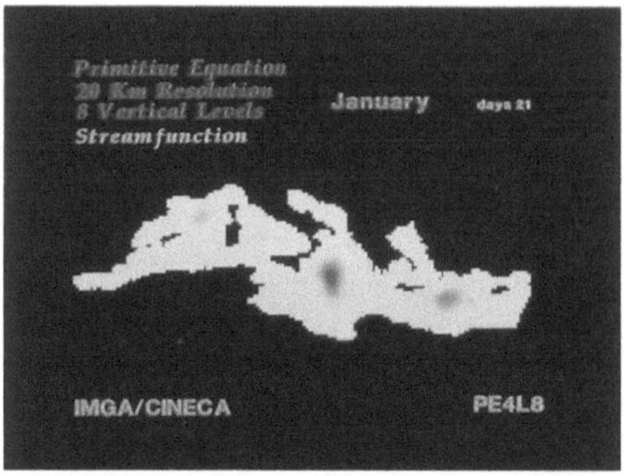

Fig. 5. The transport stream function from a wind-driven eddy resolving general circulation ocean model run on the CRAY X-MP/48. (By courtesy of N. Pinardi and A. Navarra, IMGA-CNR, Modena, Italy)

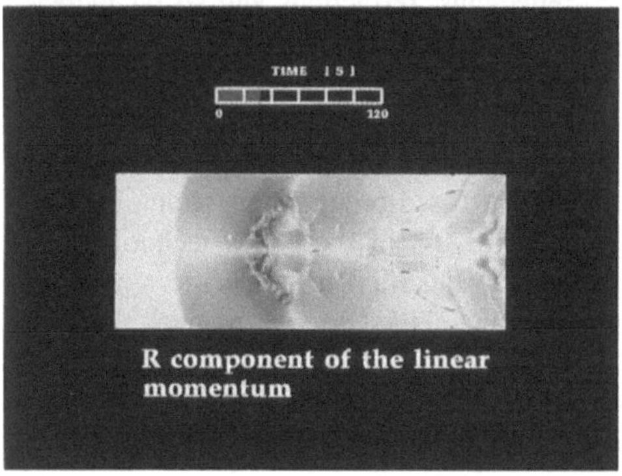

Fig. 6. A frame of a movie made at CINECA ViScLab from a numerical computation on the interaction of a strong shock with a spherical bubble. (By courtesy of R. Bedogni, Osservatorio Astronomico di Bologna, and A. Messina, Dipartimento di Astronomia, Universita' di Bologna)

Fig. 7. A three-dimensional model of a collagen fibril developed with MOVIE.BYU. (By courtesy of Istituto di Anatomia Umana Normale, Universita' di Bologna)

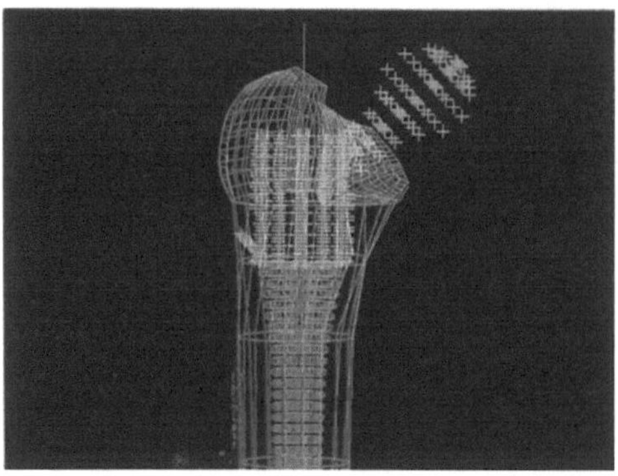

Fig. 8. A three-dimensional model of a titanium prosthesis of a femoral stem realized by Sperri (Pordenone) in cooperation with CINECA

3. Remote Scientific Visualization: An Italian Research Project

Problems in the field of scientific visualization in supercomputing environments are dealt with within research partially sponsored by the National Research Council (CNR). The research is included within the Italian Finalized Project "Information Systems and Parallel Computation" (1989-1993) in one of the themes of the first subproject called "Scientific Computation for Supercomputers".

The finalized project, which is divided into eight subprojects, has totally assigned to 148 research proposals grants which amount, for the first year (1989) to ten millions of US dollars.

The theme, containing the research, is titled "SuperComputer Access Networks". One of the aims of this theme concerns the investigation of the performance of graphical applications distributed over a high-speed network for accessing to supercomputers. This aim forecasts the experimentation of workstations specialized in the transmission of a great deal of data and in the visualization of information based on high density graphics or moving images.

Our research is included in this context. In addition, CINECA, other CNR Institutes (among them IAC – Istituto per le Applicazioni del Calcolo –), Universities and Italian Industries take part in this research. All these operative units must integrate the CNR grants with their own resources. As far as high speed network architectures for the access to a supercomputer through remote workstations is concerned, we will individuate standard functionalities to handle models of physical systems and their spatial-temporal evolution and we will analyze algorithms for realistic and dynamic visualizations [5]. We will also experiment with prototype software on application samples.

The applications we are interested in are geographically distributed over the Italian country and they concern modelling and simulation in the field of computational fluid-dynamics, computational physics and geophysics.

The complexity of systems which are dealt with in this area requires supercomputing facilities and high speed networks to transmitt large amounts of data to remote graphical workstations. General graphical requirements of these applications are interactive visualizations of scalar and vector fields in two and three dimensional space and shaded images.

The research in these fields is being carrying out within the granted groups in the Finalized Project we have mentioned above, which are actually connected to CINECA via satellite (Fig. 3).

The research carried out by CIRA aims at the numerical modelling of the dynamics of non viscous fluids in two and three dimensional domains. The images obtained by the post-processing allow the analysis of numerical simulations, the tuning of control parameters and the comparison of numerical data with experimental ones. The interactivity is mainly addressed to computation monitoring by means of graphical feedback.

The operative unit in Politecnico di Milano aims at the simulation of complex physical systems. The complexity of the systems is determined by the presence of complex structural units or by considering the quantum character of some of the degrees of freedom. The use of vector and parallel computers allows the simulation of realistic systems (up to 3000 particle systems). In particular, inside the operative unit mentioned above, there are two research groups CISCA in Trento and SISSA in Trieste, that are connected by satellite to CINECA. The CISCA group studies the simulation of atom dynamics in liquid amorphous and quasi-cristalline systems. SISSA researchers are mainly oriented to molecular dynamics problems.

Another research group at OGS (Osservatorio Geofisico Sperimentale) in Trieste, which will be soon connected to CINECA via a high speed link, is oriented to study seismic data migration in the frequency wavenumber domain and model the propagation of seismic signals. We have to point out that in the animations for migration problems the independent variable is the propagation speed of elastic waves rather than the time.

4. Conclusions

In this paper, issues concerning Scientific Visualization in supercomputer environments have been outlined.

Facilities and some applications carried out in the local area network environment of CINECA ViScLab have been examined.

The national project in this paper referred to with regard to supercomputing distributed graphics, that is partially sponsored by Italian National Research Council within one of its Finalized Projects, is now in its infancy. Actually some application areas are still being examined by means of classifying their requirements of interactivity and animation, supercomputing resources and network bandwith.

The research is aimed to individuate, and furthermore to develop, visualization functionalities that are common to different application fields. A standard approach will also be taken into account, in order to realize portable applications.

This work was performed under the auspices of CINECA and IAC, both engaged in the Italian Finalized Project "Information Systems and Parallel Computation".

5. References

1. Visualization in scientific computing. Computer Graphics *21*, no. 6 (1987)
2. G. Fabiani, M. Lanzarini, R. Rossi: Grafica e Supercomputer: il Laboratorio Grafico del CINECA. PIXEL, anno 9, no.9 (1988) 45–52
3. G. Fabiani, A. Tralli, R. Occhi et al.: Non-traditional applications at CINECA: a structural analysis of artistic works and monuments. CUG Proc., Sept. 1987, pp. 138–143
4. G. Fabiani, L. Calori et al.: Visualizzazione e analisi computerizzata del crocefisso ligneo giottesco di S. Felice in Piazza. PIXEL, anno 10, no. 1/2 (1989) 15–20
5. R. Cossu, L. Moltedo: An extension of CGI functions for generation and manipulation of raster images. Computer & Graphics *13*, no. 1 (1989) 39–49

IFIP Attendance List

John Amason
Landmark Graphics
333 Cypress Run
Houston, TX 77094, USA

Lyn Bartram
University of Waterloo
Institute for Computer Research
Waterloo, Ontario, Canada

R. Daniel Bergeron
University of New Hampshire
Kingsbury Hall -
Computer Science Department
Durham, NH 03824, USA

Steven Belcher
Liberty Mutual Insurance Co.
225 Borthwick Ave.
Portsmouth, NH 03801, USA

Christof Blum
Frauenhofer AGD
Wilhelminenstraße 7
D-6100 Darmstadt, FRG

K. S. Booth
Computer Graphics Lab
Department of Computer Science
University of Waterloo
Waterloo, Ontario, Canada NLL 3G1

Fulvio Bresciani
Aeritalia,
Strada Privata Aeroporto 10072,
Caselle, Torrino, Italy

Alexander Brown
Marathon Systems Group, Inc.
Three Main Street
Hopkinton, MA 01748, USA

B. Chanasyk
Eaton Corp.
Jon Beam Semiconductor Division
Beverly, MA, USA

W. B. Cowen
Computer Graphics Lab
Department of Computers Science
University of Waterloo
Waterloo, Ontario, Canada N2L 3G1

José L. Encarnação
Graphische Datenverarbeitung
Wilhelminenstraße 7
D-6100 Darmstadt, FRG

Georges Grinstein
Computer Science Department
University of Lowell
One University Avenue
Lowell, MA 01854, USA

G. Fabiani
CINECA, v. Magnanelli 6/3
Casalecchio di Reno
Bologna, Italy

Thomas Haaker
Frauenhofer AGD
Wilhelminenstraße 7
D-6100-Darmstadt, FRG

Naota Inamoto
Department of Information Science
University of Tokyo
7-3-1 Hongo Bunko-Ku
Tokyo 113, Japan

Hans Joseph
Frauenhofer AGD
Wilhelminenstraße 7
D-6100 Darmstadt, FRG

John F. Koegel
University of Lowell
Computer Science Dept.
One University Avenue
Lowell, MA 01854, USA

Wolfgang Krueger
Art & Com.
Hardenbergplatz 2
D-1000 Berlin 15, FRG

Fons Kuijk
Centre for Mathematics
and Computer Science
Kruislaan 413 1098 SJ Amsterdam
The Netherlands

Tosiyasu L. Kunii
Department of Information
Science, University of Tokyo
7-3-1 Hongo Bunkyo-ku
Tokyo 113, Japan

M. Lanzarini
CINECA, v. Magnanelli 6/3
Casalecchio di Reno
Bologna, Italy

Myeong Won Lee
Department of Information Science
University of Tokyo
7-3-1 Hongo Bunkyo-ku
Tokyo 113, Japan

Dwight Lillie
Chemical Abstracts Service
P.O. Box 3012
Columbus, OH 43210, USA

Augusto Mazza
Dipartimento Di Energetica
Politecnico di Torino
Corso Duca Degli Abruzzi, 24
10129 Torino, Italy

Thomas McCarken
Colorado State University
Department of Anatomy and Neurobiology
Fort Collins, CO 80523

Sean McElroy
Eaton/SED
108 Cherry Hill Drive
Beverly, MA 01915, USA

Laura Moltedo
IAC-CNR
Viale del Policlinico 137
00136 Rome, Italy

Enrico Nardelli
Instituto di Analisi dei
Sistemi ed Informatica
C.N.R., V. le Manzoni 30
00185 Rome, Italy

Paolo Dell'Olmo
Instituto di Analisi dei
Sistemi ed Informatica
C.N.R., V. le Manzoni 30
00185 Rome, Italy

Giampiero Pecelli
University of Lowell
Computer Science Dept.
One University Ave.
Lowell, MA 01854, USA

Ronald Pickett
University of Lowell
Computer Science Dept.
One University Ave.
Lowell, MA 01854, USA

Pier Paolo Rinaldi
Aeritalia – Systems and RPVs group
10072 Caselle (Torino), Italy

Toshiaki Satoh
Ricoh Company, Ltd.
Tomin-Nissei-Kasugacho-Bldg.
1-1-17, Koishikawa, Bunky-ku
Tokyo, Japan

Harald Selzer
Frauenhofer AGD
Wilhelminenstraße 7
D-6100-Darmstadt, FRG

Stuart Smith
University of Lowell
Computer Science Dept.
One University Ave.
Lowell, MA 01854, USA

T. L. Spurgeon
Departments of Anatomy,
Engineering and Mathematics,
Colorado State University
Fort Collins, CO, USA

Maurizio Talamo
Dipartimento di Informatica
e Sistematica,
Universita di Roma "La Sapienza"
Via Buonarroti 12
00185 Rome, Italy

Fabrizio Tapparo
Aeritalia - Systems and RPVs group,
10072 Caselle (Torino), Italy

Brian Thompson
IBM
Neighborhood Road Bldg. 935-933
Kingston, NY 12401, USA

Paola Vocca
Dipartimento di Informatica
e Sistematica
Universita di Roma "La Sapienza"
Via Buonarroti 12
00185 Rome, Italy

Marian Williams
University of Lowell
Computer Science Dept.
One University Ave.
Lowell, MA 01854, USA

MIX
Papier aus verantwortungsvollen Quellen
Paper from responsible sources
FSC® C105338

If you have any concerns about our products,
you can contact us on
ProductSafety@springernature.com

In case Publisher is established outside the EU,
the EU authorized representative is:
Springer Nature Customer Service Center GmbH
Europaplatz 3, 69115 Heidelberg, Germany

Printed by Libri Plureos GmbH
in Hamburg, Germany